Les Guêpes.

Two manuscript pages from Volume 8 of Fabre's *Souvenirs entomologiques. (Collection Musee de l'Harmas, Serignan)*

Fabre

INSECTS

The last photograph of Jean Henri Fabre, taken in 1914.
(Collection Kurt Guggenheim)

Nature Classics

Jean Henri Fabre

INSECTS

Edited by David Black
Illustrated by Stephen Lee

CHARLES SCRIBNER'S SONS • NEW YORK

Copyright © 1979 The Felix Gluck Press
Ltd., Twickenham, England

Library of Congress Cataloging in Publi-
cation Data
Fabre, Jean Henri Casimir, 1823-1915.
Jean Henri Fabre's insects.
1. Insects. 2. Spiders. 3. Scorpions.
I. Title.
QL467.F19 1979   595.7   78-10557
ISBN 0-684-15977-5

Printed in Germany (Federal Republic) by
Konkordia GmbH 758 Bühl/Baden

1 3 5 7 9 11 13 15 17 19 I/C 20 18 16 14 12 10 8 6 4 2

Contents

9

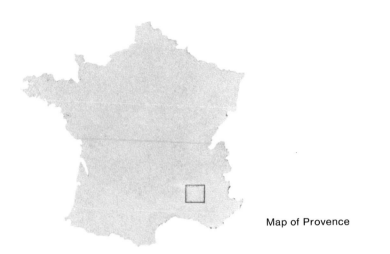

Map of Provence

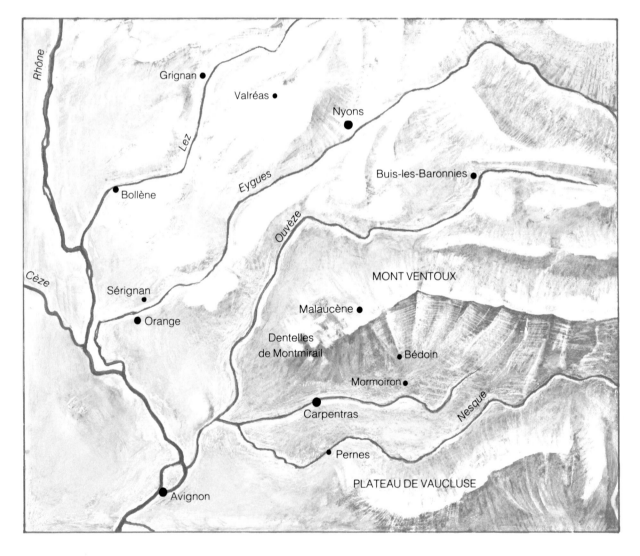

Introduction

Jean Henri Fabre died in October 1915 at his country retreat in Vaucluse as the Great War was sweeping through Northern Europe. He was ninety-one. The death of one of France's masters passed almost unnoticed. His world was not the world of the celebrated academic scientists but the intimate world of the insects he loved.

Fabre was born in 1823 at St Léons, a hamlet in the province of Aveyron on the southern flanks of the Massif Central. His parents were poor, uneducated peasants. The legacy of their poverty dogged him through his early years, and the first half of his life was an uphill struggle.

As a small boy, he was a slow learner, a dreamer who showed little interest in conventional studies. One day, however, his father brought home a poster on which the alphabet was portrayed by the first letters of animal names together with pictures of the animals. Fabre was fascinated, and within a few days had learned all the letters of the alphabet. 'Those speaking pictures, which brought me among my friends the animals, were in tune with my instincts.'

To supplement their earnings, his parents decided to keep ducks. The night young Fabre was told that he was to be the duck boy, he became so excited that he dreamed he was in Paradise leading his fluffy ducklings to the pool. A month later his dream was realized and he led his twenty-four ducklings up into the hills to a little stream that widened out into a pond. There, leaving his charges to amuse themselves, he focused his attention on the life of the pond — its mosses, tadpoles, shells, beetles and the broken stones that glinted like diamonds. Returning home at the end of the day with his pockets bulging with curios, he was sharply reprimanded for collecting so much rubbish and made to throw it away. But he had discovered his lifelong passion.

When he was twelve, his parents moved from the country to the town of Rodez, where his father kept a café. School for Fabre was enlivened by Virgil's descriptions of nature : 'exquisite details concerning the cicada, the goat and the laburnum tree'.

Over the next five years the family was constantly on the move. Fabre was soon forced to leave home to earn a living and he turned his hand to whatever job he could find — selling lemons in the marketplace or working on the railroad. Yet, despite the harshness of his existence, Fabre was able to find solace in nature : he continued to follow the lives of wasp and beetle in every spare moment he had.

In 1841, when he was just eighteen, Fabre won a scholarship to the Ecole Normale Primaire, the teachers' training college at Avignon, and for the next thirty years and more his life was dominated by schools and classrooms. He achieved a fine record as a student and on leaving the college was offered the post of schoolmaster at nearby Carpentras, where he met and married Marie Villard. There, when he could get away from teaching reading, writing and elementary chemistry to unruly schoolboys, he wandered the fields, enjoying the exercise and the opportunity to increase his intimacy with nature.

He also kept up his studies, took more exams, and in 1850 accepted the post of lecturer at the college of Ajaccio in Corsica. He was charmed and intoxicated by the island's rugged landscape, deep woods and mountains rich with scented flowers. Here, amongst a profusion of plants and animals, his mind was greatly stimulated and all the scientific knowledge he had gained so far seemed to come together.

His stay on Corsica sadly lasted only three years. He was struck down by malaria, probably contracted during his nocturnal roamings along the marsh-backed beaches. The disease left him so weak that he had to return to the mainland, where he was offered the post of science teacher at the high school in Avignon.

In 1855 he wrote his memoir on the hunting wasp Cerceris, an early demonstration of his supreme gifts both as observer and recorder of insect life. He was now thirty-two years old with five children to support, and he had to supplement his meagre income by giving private lessons. In the same year he submitted papers for a doctorate in natural sciences. Through this not particularly original work on the structure of wild orchids and the reproductive organs of centipedes and millipedes he gained his 'piece of paper', and returning to Avignon soon embarked on a study of a group of fungi, Spheriacae, a project he was engaged on for a further twenty years.

But Fabre's true hobby remained insects. Around Avignon he discovered many fresh localities. A sandy plateau at Les Angles, just across the Rhône, served as a stage where he could watch sacred scarabs roll their dung. Another favourite spot was Issarts wood, a dense area of ilex scrub, and there he observed the hunting wasp Bembix in the quest for flies. The outskirts of Carpentras were good spots, for the soft marl — a kind of limestone mixed with sand and clay — was ideal soil in which the burrowing Hymenoptera could establish their dens.

The satisfaction he derived from the countryside and from his family provided him with a much needed antidote to the stifling atmosphere of the teaching profession in Avignon. He gave free lectures in the abbey of

13

Saint-Martial but, though he was popular, his ideas of teaching girls science were viewed as dangerous and subversive.

He crept up the teaching ladder, and was eventually offered the extra duty of drawing master and appointed part-time curator of the local museum. But these perks did little to boost his finances. Desperate to make more money, Fabre spent years developing a simple method of extracting dye from madder, but he had little business sense and unwittingly revealed his technique to men whose sharper sense enabled them to profit from his hard labours and sabotage his work. It was a bitter blow.

Fabre was victimized by his colleagues, and in 1871, in what can only be described as a witch hunt, he and his family were ousted from their lodgings. On top of all this he fell ill. 'With the little lucidity I had left to me, and having nothing else to observe, I watched myself dying. I observed with a certain degree of interest the gradual falling apart of my machinery.'

His recovery was in part due to a chance acquaintance with the English philosopher John Stuart Mill, who had lived in Avignon and accompanied Fabre on walks. Fabre wrote to him in London, explaining his grave circumstances. Mill replied at once, and sent £120 with a view to freeing Fabre from all ties with Avignon.

In the event, Fabre settled not very far away, on the outskirts of Orange, in a pleasant house situated in a great meadow, with a fine avenue of plane trees connecting it to the small road to Camaret. It was here that he wrote a number of textbooks for young people, aimed at stimulating their interest and enthusiasm for the sciences. He wrote on earthquakes and mushrooms, volcanoes, sun and thunder, butterflies and bees. These books not only provided Fabre's livelihood for nine years, but proved good practice for the more serious work that was to come.

One day, for no apparent reason, the landlord decided to cut down the avenue of plane trees. To Fabre, yearning for peace and quiet, this simple act was enough to make him finally decide to abandon the town for ever and find a place of his own. He chose a house in Sérignan, a village in Vaucluse, and it was here that, at the age of sixty, he started a new life. His wife had died, and Fabre married again, a younger woman who bore him three children. He was content in his country retreat, surrounded by his family and the stone walls of his harmas. Previous skirmishes with fellow scientists had taught him not to waste time on discussion and correspondence, and his days were joyfully occupied studying his insects while the seasons raced by. For thirty full years he was able to devote himself to the study and close observation of the creatures he loved. The

Experimental apparatus filled with earth for studying the burrowing activities of dung beetles.

14

A box with protruding wooden tubes of varying lengths to mimic the dried hollow stems of plants. This entices osmia bees to lay their eggs within each tube and enables controlled experiments to be made.

fruit of all his work was Souvenirs entomologiques, his ten-volume treatise on the behaviour of insects.

These masterpieces of scientific discovery (published from 1879 to 1907) cover many insect groups. Among his most famous studies are his work on hunting wasps, his descriptions of the domestic lives of dung beetles, the revelation of the complicated life cycle of a group of beetles called Meloidae (for a part of their life cycle Fabre coined the term hypermetamorphosis), and detailed work on the relationship between the sex of the egg and the dimension of the cells of a group of solitary bees, Osmiae.

But despite all this publishing output, Fabre's financial worries were not over. As late as 1908 he was reduced to poverty. He tried to sell his beautiful watercolours of fungi. A plea to the poet Mistral brought help and also saved his watercolours; and later the local council gave him a small retainer. In 1910, his village paid their respects with a celebration in honour of their scientist, a celebration which sadly lacked the participation of fellow scientists. However, during the last years of his life the popularity of the Souvenirs increased and, in the few short years remaining, the books could be counted a success.

In the end, what is to be said of Fabre? His expansive style of writing runs counter to the work of many other scientists. Yet, reading Fabre, one is immediately struck by the man's amazing ability to paint pictures with words, to bring his insects to life, creeping and crawling beyond the pages of the book. At times he can be as gripping as a great novelist. In common with many other largely self-educated people, he shows much originality of thought, and to those who follow him, he leads the way into many bypaths, striking out along unbeaten tracks. Some of his work can be criticized. Although his botanical work and herbarium can hardly be faulted in terms of description or classification, he was in no way a strict taxonomist or anatomist; he was far more interested in behavioural studies. He saw emotional relationships among insects where none exist, and he paid scant regard to Darwin, for he was convinced of the fixed nature of all species. However, there is no doubt that Fabre was one of the great observers of nature and has left us with some of the most exciting and vivid accounts of insect life.

In extracting brief texts from the huge volume of Fabre's work, we have generally grouped the insects together according to type, and in so doing have ignored the chronological order of his studies. Our aim has been both to offer a taste of his evocative and fascinating writings, and also to stimulate the reader, with the help of the illustrations, to take a fresh look for himself at the amazing world of insects.

The Harmas

This is what I always wished for, a bit of land, not very large, but fenced in to avoid the drawbacks of a public way; an abandoned, barren, sun-scorched bit of land, favoured by thistles and wasps and bees. For forty years I have fought against the paltry plagues of life for this long wished for laboratory.

It is a little late, my pretty insects. I greatly fear that the peach is offered to me when I am beginning to have no teeth to eat it!

My harmas abounds with the invaders of any soil that is first dug up and then left for a long time to its own resources — the yellow-flowered centaury, the star thistle and the rough centaury. Here and there amid confusion stands the Spanish oyster plant whose spikes are as strong as nails. Above it towers the cotton thistle, whose straight and solitary stalk soars to a height of one or two metres and ends in large pink tufts. To visit this prickly thicket when the wasp goes foraging, you must wear boots that come to mid-leg or else resign yourself to smarting calves. Into my Eden — cursed ground to others — come the bees and wasps. Its mighty growth of thistles and centauries draws them all to me from everywhere around. Never have I seen so many gathered in a single spot. Here come the hunters of every kind of game, builders in clay, weavers, architects in pasteboard, carpenters and miners.

16

The pink house with green shutters lies half hidden at the end of an avenue of lilac trees. Since Fabre's death in 1915 it has remained much as he left it—a museum for all interested in natural history. Upstairs his study facing the garden holds a large collection of insects, fossils, shells and pressed plants, while downstairs there is a gallery the walls of which are lined with his delicate water-colours of wild mushrooms. Outside in the garden, beyond the famous plane trees, is the secluded pool, and beyond this a selection of the wild plants of Provence.

I have seen the Ammophilae fluttering along the garden walks in search of a caterpillar; the Pompili, who travel alertly, beating their wings and rummaging in every corner in quest of a spider. The largest of these wasps waylays the Narbonne *Lycosa*, whose burrow is not infrequent in the harmas. This burrow is a vertical well, with a curb of fescue grass intertwined with silk. You can see the eyes of the mighty spider gleam at the bottom of the den like little diamonds. And here on a hot summer afternoon, is the Amazon ant who leaves her barrack rooms in long battalions and marches far afield to hunt for slaves. Here again, around a heap of grasses turned to mould, are the Scoliae, four centimetres long, who fly gracefully and dive into the heap, attracted by a rich prey, the grubs of the biggest of our beetles.

What subjects of study! And these were not all. The house was as utterly deserted as the ground. With man gone and peace

European tree frog (*Hyla arborea*)

assured, the animals hastily seized on everything. The warbler took up residence in the lilac shrubs, the greenfinch settled in the thick shelter of the cypress. The serin finch, whose downy nest is no bigger than half an apricot, came and chirped in the plane tree tops, while in the night the scops owl took up position.

In front of the house is a large pond. Here, from half a mile and more around, come the frogs and toads in springtime.

The natterjack, sometimes as large as a plate, with a narrow stripe of yellow down its back, comes to take a bath. In early evening we see the midwife toad hopping along the edge, the male carrying a cluster of eggs the size of peppercorns entwined around his hind-legs. He has brought his precious packet from afar to leave it in the water before retiring under some flat stone, whence he emits a sound like a tinkling bell. Lastly, when not croaking amid the foliage, the tree frogs indulge in the most graceful dives. And so in May, as soon as it is dark, the pond becomes a deafening orchestra; it is impossible to talk, impossible to sleep.

Bolder still, the wasp has taken possession of my house. On my doorstep in a soil of rubbish nests the white-banded *Sphex*. It is a quarter of a century since I last saw this saucy cricket-hunter. On the moulding of the venetian blinds, a few stray mason bees build their group of cells; inside, a mason wasp builds her earthen dome. The common wasps and paper wasps are my dinner guests; they visit my table to see if the grapes served are as ripe as they look.

Natterjack toad (*Bufo calamita*)

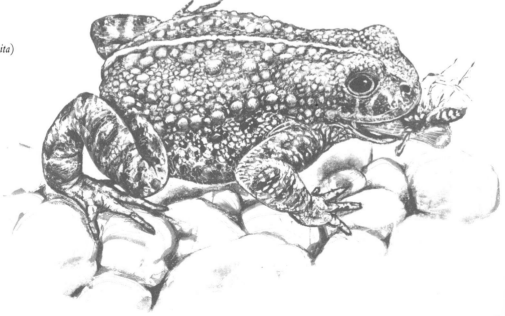

Scarce swallowtail butterfly (*Iphiclides podalirius*)

Here surely, and the list is far from complete, is a company both numerous and select, whose conversation will not fail to charm my solitude. Besides, should I wish for even more variety, Mont Ventoux is but a few hundred steps away, with its tangle of strawberry trees, rock-roses and heather. That is why, foreseeing these riches, I have abandoned the town for the village and come to Sérignan to weed my turnips and water my lettuces.

Laboratories are being founded at great expense along all our coasts, where people cut up small sea animals of meagre interest. They spend a fortune on powerful microscopes, dissecting instruments, boats, fishing crews, to find out how the yolk of a sea worm's egg is constructed, and they scorn the little land animal which lives in constant touch with us. One might think it would be more important to gain a knowledge of the life history of the destroyer of our vines than to know the nerves of a barnacle. The depths of the sea are explored with many drag nets; the soil that we tread is disregarded. While waiting for the fashion to change, I open my harmas laboratory of living entomology; and this laboratory will not cost the ratepayer one penny.

19

The Ant and the Cicada

Fame is built mainly on legend; in the animal world as in the world of men, the story takes precedence over history.

Take the cicada: memorable verses describe her first when the bitter winds begin to blow; quite destitute she hurries to her neighbour the ant. But when she complains to the ant that she is hungry, she meets with but a poor reception.

> Vous chantiez! J'en suis bien en aise.
> Eh bien, dansez maintenant!

> (You used to sing! I'm glad to know it.
> Well, try dancing for a change!)

These two short lines enter the child's mind like a wedge and never leave it. The truth is, however, just the reverse.

Cicadas (*Lyristes plebjus*), are large bugs related to aphids; most live in warm climates, though there is one exceedingly rare British species. Species found in France are brown or black and measure 16–47 mm; they are smaller than the tropical varieties, which are often brightly coloured red and violet.

In July, during the stifling heat of the afternoon, when the insect population vainly wanders round the limp and withered flowers in search of refreshment, the cicada laughs at the general need. With that delicate gimlet, its rostrum, it bores through the firm, smooth bark swollen with sap ripened by the sun. Driving its sucker through the bung-hole, it drinks luxuriously, motionless, absorbed in the charms of syrup and song.

Watch for a little while. There are many thirsty ones prowling around; they discover the well, betrayed by the sap that oozes from the margin. The smallest, in order to reach the well, slip under the abdomen of the cicada, who good-naturedly raises itself on its legs and leaves a free passage for the intruders. The larger ones quickly snatch a sip, retreat, take a walk and then return to show greater enterprise.

Of all these insects, the worst offenders are the ants. I have seen them nibbling at the ends of the cicada's legs; I have caught them tugging at the tips of its wings, tickling its antennae and even trying to pull the sucker out.

Worried by these pygmies, the giant ends by abandoning the well. It flees, spraying the robbers with urine as it goes.

Cicada attacked by ants.

21

The Cicada—a Life History

Emergence

The first cicadas appear just before midsummer. Along the much trodden paths baked by the sun, there open, level with the ground, round holes about the size of a man's thumb. These are the exit holes of the cicada larvae, which come up from the depths to undergo their transformation on the surface. Equipped with powerful tools to pass if necessary through sandstone and hard clay, the larva on leaving the earth has a fancy for the hardest places.

The cylindrical tunnel runs perpendicular for about 40 centimetres, ending in a rather wider chamber. The larva can move nearly up to the surface and down again without producing landslips, for it is quite a clever engineer, and cements its shaft.

Transformation

The exit-gate is passed and left wide open, and for some time the larva wanders about looking for some aerial support—a tiny bush or a tuft of thyme. It finds it, climbs up and, head upwards, clings on firmly with the claws of its fore-feet. First the thorax splits to reveal the pale green colour of the insect. Then the wrapper of the skull breaks crosswise in front of the eyes, and the red stemmata appear. We see slow palpitations, alternate contractions and distensions caused by the ebb and flow of blood. The skinning operation makes rapid progress. Soon the head is free, then the front legs and finally the back ones. This first transformation takes ten minutes; the second is much longer and involves the sloughing off of the skin from the abdomen. Some three hours later the green cicada changes to brown before my eyes, and flies away from its nursery branch.

In Europe, adult cicadas live for only 4 to 6 weeks, but the larvae take 4 years to develop. By patient observation, Fabre discovered the cicada's method of egg laying and the development of the strange primary larva.

22

Eggs and Development

Two to three weeks after her emergence, that is to say about the middle of July, the cicada busies herself with her eggs. Small dry branches are those most favoured, especially the sprigs of Spanish broom, which are like straws crammed with pith. The cicada disposes of three to four hundred eggs. Very often, while she is absorbed in egg-laying, an infinitesimal wasp, herself the bearer of a boring tool, labours to exterminate the eggs as fast as they are placed. The wasp stands immediately behind the cicada and by the time the mother has exhausted her ovaries and flies away, most of her cells have been parasitized by the alien egg.

Above: underside of cicada showing two plates beneath which tendons vibrate to produce a loud drumming sound.

Below: cicada laying eggs. Nymphs lowering themselves to the ground by silk threads.

The Larvae

By early October, there appear on the eggs two little dark brown spots, round and clearly defined. These two shining eyes, which seem to look at you, combined with the cone-shaped fore-end, give the eggs the appearance of finless fishes, the very tiniest of fishes. I collect in boxes, tubes and jars a hundred twigs of all sorts colonized with these eggs; not one of them shows me what I want to see: the emergence of the budding cicada. At last on 27 October, despairing of success, I propose once more to examine the cells and their contents. The first fire of the season has been lit. I put my bundle on the chair before the hearth, without any intention of trying the hot flames on the nests. While I am passing my magnifying glass over a slit stem, the hatching suddenly takes place. My bundle comes alive, and the young larvae emerge from their cells by the dozen.

In its general shape, the configuration of the head, and the large black eyes, the creature, even more than the egg, resembles a fish. A mock ventral fin accentuates the likeness. But this state, the primary larva, does not last long. The true larva soon emerges and, swinging by a thread positioned from its abdo-men, it gently sways and lowers itself to the ground. I know of hardly any more curious sight than that of this miniature gymnast hanging by its hinder part, making ready in the air for its somersault into the world.

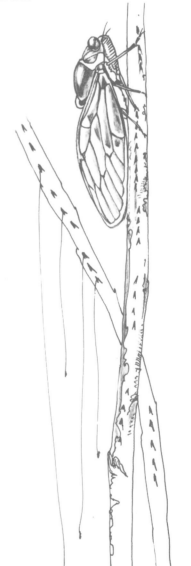

23

The Night of the Great Peacock
(Emperor Moth)

It was a memorable evening. I shall call it the night of the great peacock. Who does not know this magnificent moth, the largest in Europe, clad in maroon velvet with a necktie of white fur? The wings sprinkled with grey and brown are crossed by a faint zigzag and have in their centre a round patch, a great eye with a black pupil and iris containing successive black, white, chestnut and purple arcs.

No less remarkable is the caterpillar. An undecided yellow in colour, it has on its top a thin scattering of tubercles, crowned with a palisade of black hairs set with beads of turquoise blue. Its stout brown cocoon is usually fastened to the bark at the base of old almond trees. The caterpillar feeds on the leaves of the same tree.

On the morning of 6 May, a female emerges from her cocoon on the table of my insect laboratory. Just to watch — the habit of the observer, always on the look-out for what may happen — I keep her captive under a wire gauze bell-jar.

Fabre conducted his studies on the great peacock moth, more commonly known as the greater emperor (*Saturnia pyri*). This is the largest European moth, with a wingspan of 15 cm. Fabre carried out similar experiments on the smaller emperor moth (*Saturnia pavonia*) and on the handsome oak eggar moth (*Lasiocampa quercus*).

24

What a lucky thought!

At nine o'clock in the evening, just as the household is going to bed, there is a great stir in the room next to mine. Little Paul half-undressed is rushing about, jumping, stamping and knocking the chairs over like a mad thing.

'Come quick!' he screams. 'Come and see these moths, big as birds! The room is full of them!'

I hurry in, to find that there is ample justification for the child's enthusiastic outcry—an invasion as yet unprecedented in our house, a raid of giant moths. We run downstairs.

We enter my study, candle in hand. What we see is unforgettable. With a soft flick-flack, the great moths fly around the bell-jar where only this morning a female peacock had emerged from her cocoon. They rush at the candle, putting it out with a stroke of their wings; they descend on our shoulders, clinging to our clothes, grazing our faces. The scene suggests a wizard's cave; there must be forty or more in the house.

The weather is stormy; the sky is overcast and the darkness so profound that even in the open air, far from the shadow of the trees, it is hard to see one's hand before one's face.

Whenever the night is pitch dark, the moths arrive between eight and ten o'clock. They come one by one and keep coming as long as the female moth lies captive in my study. The house is hidden by tall plane trees, clumps of pine and screens of cypresses. Just how, through this tangle, does our moth reach the object of his pilgrimage?

Below left: greater emperor caterpillar. Below right: cocoon cut open to show chrysalis.

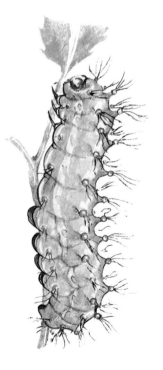

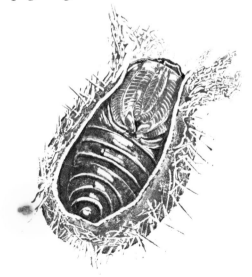

The Secret of the Emperor Moth

Fabre was completely baffled by this invasion. To understand how and why it happened he embarked on a series of experiments that took three years to perfect, owing to difficulties in finding and rearing the cocoons and also to the changeable weather of early May when the moths are on the wing. When at last his females did emerge, he tried by a process of elimination to determine the factors involved. His efforts included hiding the female out of sight of the males, moving her to different places in his study, and filling the study full of strong-smelling substances such as naphthalene. None of these had any effect. It was only when he cut off the males' large plumed antennae and held the female captive in an airtight bell-jar that the invasion ceased. From this he deduced that the female's scent was the all-important factor, but even then he still believed that unknown radiations from the female guided the males over great distances.

The year of his crucial experiment was 1879. It was not until seventy years later that the first insect sex attractant was isolated, for the silk moth, by workers at the Max Planck Institute in Germany. Other attractants —pheromones— have since been found, including that for the gypsy moth, which is used in North America to control the spread of this harmful insect.

Left: the female emits pheromones from her abdomen.
Right: the male detects the pheromones by means of his large antennae.

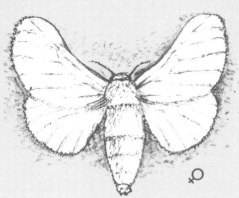

We already know a lot more about the structure and function of pheromones. They are extremely potent in minute quantities: male moths can be attracted over a distance of two to three miles by a single female. Moth pheromones are produced by a pair of glands located between the last two segments of the female's abdomen. These glands are protruded from the body when the moth is ready to receive males. The attracted males circle the female with their abdomens bent towards her; at the same time the males' wings beat rapidly in what is called the whirring dance. In many species the male moth in turn excites the female by producing a scent from glands under his wings.

Chemical formula of silk moth pheromone.

$$
\begin{array}{ccccccccccccccccc}
 & H & H & H & H & H & H & H & H & H & H & H & H & H & H & H \\
 & | & | & | & | & | & | & | & | & | & | & | & | & | & | & | \\
H- & C- & C- & C- & C= & C- & C= & C- & C- & C- & C- & C- & C- & C- & C- & C- & C-OH \\
 & | & | & | & & & & | & | & | & | & | & | & | & | & | \\
 & H & H & H & & & & H & H & H & H & H & H & H & H & H \\
\end{array}
$$

Other insects produce pheromones for different purposes: ants produce several types to arouse various responses in members of their own colony. Cockroaches and sawflies are just two other types that utilize the sex attractant type of pheromone.

Other moths that use pheromones: above, male and female silk moths (*Bombyx mori*); below, male and female gypsy moths (*Porthetria dispar*).

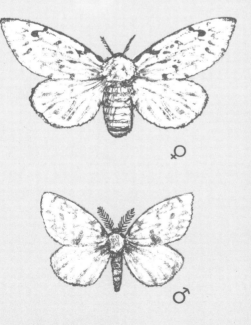

Pine Processionary Caterpillars

The Eggs and The Hatching

In my harmas laboratory stand some vigorous fir trees. Every year the caterpillar takes possession of them and spins his great purses in their branches. You voracious creatures, if I let you have your way, I should soon be robbed of the murmur of my once leafy pines. Let me make a pact with you. You have a story to tell. Tell it to me; and for a year, for two years, or longer, until I know more or less all of it, I shall leave you undisturbed.

In the first two weeks of August, we discover little whitish cylinders spotting the dark green. These are the moths' eggs: each cylinder is the cluster laid by one mother. This muff, which has a silky appearance, is covered with scales that overlap like tiles on a roof.

The hatching takes place a month later in September. The little caterpillars leave the egg in the morning at about eight o'clock. They measure a millimetre in length and are pale yellow and bristling with hairs, some shortish and black, others rather longer and white. After roaming for a few moments at random among the scales of the common cradle, most make for the leaves, which are eaten into faint and narrow grooves, only the veins being left intact.

Pine processionary (*Thaumetopoea pityo-campa*): *1* eggs and young caterpillar; *2* temporary nest; *3* winter nest.

Right: caterpillars descending from cocoon in springtime.

28

From time to time, three or four who have eaten their fill fall into line and walk in step, but soon separate. This is practice for the coming processions. When the sun reaches their corner of the window, the little family retreats to the base of the double leaf, gathers into an irregular group and begins to spin. Under this tent, a very wide-meshed net, a siesta is taken during the hottest and brightest part of the day. In the afternoon, the flock leaves its shelter, disperses and resumes feeding.

And so in less than an hour from hatching, the caterpillar is both a processionary and a spinner. In a few weeks' time a first moult replaces the humble fleece. The caterpillar is now about two centimetres, adorned with patches of bright red and palisades of scarlet bristles.

The Nest and the Community

November arrives, bringing cold weather; the time has come to build the stout winter cocoon. High up in the pine the tip of the bough is chosen, with its suitably close-packed leaves. The spinners surround it with a spreading network, which bends adjacent leaves a little nearer and ends by incorporating them into the fabric. In this way they make an enclosure, half silk, half leaves, capable of withstanding the winds and rains of winter.

The 'nest' is roughly egg-shaped, tapering below and extending into a sheath which envelops the supporting branch. Every evening between seven and nine o'clock, weather permitting, the caterpillars leave the nest and go down to the bare bough, which forms the pole of the tent. They divide into squads and disperse in every direction, all the time working their spinnerets and strengthening the axis of their winter retreat.

At the top of the actual dome are round openings, as wide across as an ordinary pencil. These are the doors of the house, through which the caterpillars go in and out. All around the shell are projecting leaves, which the insects' teeth have spared. From the tip of each leaf there radiate threads forming a spacious verandah. Here is a broad terrace on which, in the daytime, the caterpillars come and doze in the sun.

29

The Procession

The pine caterpillar is sheep-like: where the first goes all the others go, in a regular string, with not an empty space between them. We can give a full picture of its character if we add that it is a rope dancer all its life: it walks only on the tightrope, a silken rail placed in position as it advances. As the first dribbles its thread, so a second steps on the slender footboard and doubles it with its thread and so the trail thickens and strengthens all along the line.

The Moth

When March comes, the final exodus begins. After a couple of hours of marching and countermarching, the fragmentary processions, each comprising a score of caterpillars, reach the foot of a wall. Here the soil is powdery, very dry and easy to burrow in. At last some spot is recognized. The leading caterpillar halts, pushes with his head, digs with his mandibles. Then the file breaks up; the 'worm' has chopped itself into a gang of independent workers. An excavation is formed in which little by little the caterpillars bury themselves. For some time the undermined soil cracks and rises and covers itself with little molehills; then all is still.

A fortnight later, under the soil, we find the cocoons assembled in bunches — cocoons of sorry appearance, soiled with earth particles; narrow ellipsoids, pointed at both ends, measuring just under three centimetres in length. The fragility of the walls is remarkable when we have seen the enormous quantity of silk expended in the construction of the nest.

The moth appears at the end of July or in August. Its costume is a modest one: upper wings grey, striped with a few crinkly brown streaks; thorax covered with thick grey fur; abdomen clad in bright russet velvet. The last segment has a pale gold sheen. In the place of hairs it has scales so well assembled and so close together that the whole seems to form a continuous block, like a nugget. When we rub this, a multitude of scales come off and flutter shining like mica spangles. This is the golden fleece with which the mother covers the cylinder of her eggs.

The life cycle

The young hatch out from their cylindrical egg case in September. They start eating immediately, stripping pine trees of their needles. After a while they move down the branch and across to a neighbouring bough. In this way they gradually make their way up the tree. They feed and travel at night; during the day they rest in temporary nests which are spun between the branches. In November they start building their stout winter nests, from which they make nightly forays. It is during the winter months that this insect is most destructive. In March the caterpillars leave their trees and bury themselves in loose earth, where they pupate. The moth emerges in July.

Methods of control

These include mechanical removal of nests, spraying with insecticides, biological control by means of viruses and bacteria. L'Institut National de Recherche Agronomique, based near the town of Malaucène, has done much valuable research on the control of the pine processionary in the pine forests of nearby Mt Ventoux, a mountain well known to Fabre. The pine processionary is also controlled naturally by a large number of parasitic and predatory insects that attack it at all stages in its life history.

The pine moth is a very bad flyer; she flutters about and blunders to earth again, and her best efforts barely succeed in bringing her to the lower branches of the pines. Here, at a height of two metres at most are deposited the cylinders of eggs. It is the young caterpillars who, from one provisional encampment to another, gradually ascend to the summits on which they weave their final dwellings.

On 30 January 1896 Fabre led his caterpillars to the rim of an earthenware pot, then cut off their silken trail to see for how long they would go round and round. 'Each caterpillar is preceded by another on whose heel he follows. At every circuit the silken thread gets thicker; this headless file has no liberty left, no will, it has become clockwork. And this continues for hours and hours. I stand amazed at it, or rather stupefied.' It took the caterpillars eight days to find their way off the pot and home, forced by hunger and a series of accidents after one caterpillar stumbled over the rim, starting the exit to freedom. 'I am already familiar with the abysmal stupidity of insects as a class whenever the least accident occurs.'

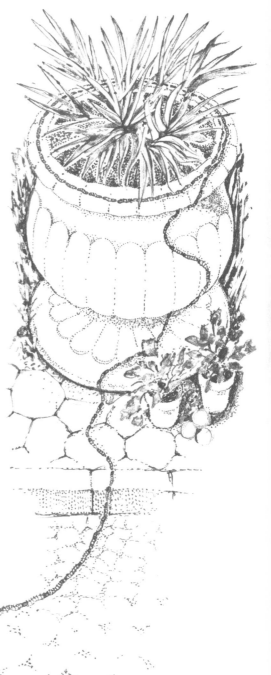

The caterpillar of the pine processionary moth is a serious pest of pine and cedar trees in countries bordering the Mediterranean. The life of the insect has a morbid fascination. Fabre takes up its story from another famous zoologist, Réaumur. Fabre reared his insects on pines that grew naturally in his harmas. He discovered their nocturnal wanderings, their nest-making procedure, their sensitivity to climate and the irritating property of their stinging hairs. He observed the minute details of their processionary behaviour from the time it develops after the last larval moult — the most damaging stage for pine trees.

Fabre's account remains a classic of observation and is used today by scientists in their fight to control this most destructive insect.

Purple saxifrage (*Saxifraga oppositifolia*)

An Ascent of Mont Ventoux

Thanks to its isolated position, which leaves it freely exposed on every side; thanks to its height, which makes it the topmost point in France within the frontiers of the Alps and the Pyrenees, our bare Provençal mountain, Mont Ventoux, lends itself remarkably well to the study of the climatic distribution of plants. Half a day's journey in an upward direction brings before our eyes such a succession of types of vegetation as we would find in the course of a long journey north through Europe to the Arctic Circle.

In the hedges below, you can pick the scarlet flowers of the pomegranate, a lover of African skies; above, you will find a shaggy little poppy as common in Greenland as on the upper slopes of Ventoux. These contrasts always have something fresh and stimulating about them.

I made my twenty-third ascent in the summer of 1865. There were eight of us: three botanists and five others attracted by a mountain expedition and the panorama of the heights. We rose long before dawn. As the sky was growing light we climbed the rocky paths. First oak and ilex disappeared by degrees; then the vine and almond tree; and next the mulberry, the walnut tree and the white oak were left behind. Box became plentiful and then came a monotonous region where winter savoury was the dominant plant. Certain small cheeses forming part of our provisions are powdered with this strong spice. Already more than one of us is biting into them in his imagination and casting hungry glances at the provision bags carried by our mules. I show my colleagues a little sorrel plant, with arrow-head leaves,

Sorrel (*Rumex scutatus*)

Pomegranate (*Punica granatum*)

32

Apollo butterfly (*Parnassius apollo*)

Alpine poppy (*Papaver rhaeticum*)

which will help to stay their rumbling stomachs until they reach the next halt.

While chewing the bitter leaves, we come to the beeches. At first these are big solitary bushes trailing on the ground; soon after, dwarf trees clustering together; and finally, mighty trunks forming a dense and gloomy forest whose soil is a mass of rough limestone blocks. We spend an hour or more crossing this wooded zone, then once more the beeches become bushy and scattered.

During the climb to the summit, some of us feel a sort of seasickness, caused by fatigue and the rarefaction of the air. At last we are there and take refuge in the rustic chapel of Sainte-Croix. Soon the sun rises. To the south and west stretch misty plains. To the north and east, under our feet, lies an enormous bank of clouds, a sort of ocean of cotton wool. A few tops, their glaciers trailing, gleam in the direction of the Alps.

Longhorn beetle (*Rosalia alpina*) on beech tree (*Fagus sylvatica*)

But botany cries out for our attention; the top of Ventoux in July is a literal bed of flowers: my memory recalls those elegant tufts of *Androsace villosa*, with its pink-centred blooms all streaming with the morning dew; the Mont Cenis violet, spreading its great blue blossoms over the chips of limestone; the spikenard valerian; the wedge-shaped globularia, forming close carpets of bright green dotted with blue; the alpine forget-me-not; the Candolla candytuft, whose tiny stalk bears a dense head of little white flowers. Here too, the opposite-leaved saxifrage with its purple flowers and the musky saxifrage with white-yellow flowers. When the sun's rays are hotter, we shall see fluttering idly from one tuft of blossom to another a magnificent butterfly, *Parnassius apollo*, its white wings adorned with four bright crimson spots surrounded by black. Its caterpillar feeds itself on saxifrage.

Night Song

We are in the middle of July; the astronomical dog-days are just beginning, but in reality the torrid season has anticipated the calendar and for some weeks past the heat has been overpowering. This evening in the village they are celebrating the national festival. While the little boys and girls are hopping around a bonfire, I am sitting in a dark corner, in the comparative coolness, harking to the night sounds of the fields, grander by far than that which at this moment is being celebrated in the village square with gunpowder, lighted torches and Chinese lanterns.

It is late; and the cicada is silent. Glutted with light and heat, it has indulged in symphonies all the livelong day. In the dense branches of the plane trees a sudden sound rings out like a cry of anguish, strident and short. It is the desperate wail of the cicada, surprised in his quietude by the green grasshopper, the ardent nocturnal huntress, who springs upon him, grips him in the side, opens and ransacks his abdomen. An orgy of music followed by butchery.

Hard by the place of slaughter, in the green bushes, a delicate ear perceives the hum of the grasshoppers. It is the sort of noise that a spinning wheel makes, a very unobtrusive sound, a vague rustle of dry membranes rubbed together. Above this dull bass there rises at intervals a hurried, very shrill, almost metallic clicking. For all that it is a poor concert, very poor indeed, although there must be ten little fellows strumming near me. My old eardrum is not always capable of perceiving these subtleties of sound.

You will never equal your neighbour, the little bell-ringing toad, who goes tinkling all round at the foot of the plane trees while you click above. On this evening of rejoicing, there are nearly a dozen of his kind tinkling against one another around me. Most of them are crouching among the rows of flower pots that form a sort of lobby outside my house, Each has his own note, always the same, lower in one case, higher in another, a short, clear note, melodious and of exquisite purity.

34

Among the July singers, one other, were he able to vary his notes, could vie with the toad's harmonious bells. This is the little scops owl with round golden eyes. He sports on his forehead two small feathered horns, which have won him the name of 'horned owl' in my neighbourhood. His song, which is rich enough to fill by itself the still night air, is of a nerve-shattering monotony. For hours on end, 'kew, kew', the bird spits out its cantata to the moon. In contrast with his soft note, there comes intermittently from another spot a sort of cat's mew. This is the call of the tawny owl. After hiding all day in the seclusion of a hollow olive tree, he starts his wanderings when the shades of evening begin to fall. Swinging along in a sinuous flight, he has come from somewhere in the neighbourhood to the pines in my enclosure, whence he mingles his harsh mewing, slightly softened by distance, with the general concert.

It is another insect that takes the prize amongst all these choristers. I speak of the pale and slender Italian cricket, who is so puny that you dare not take him up for fear of crushing him. He makes music everywhere among the rosemary bushes, while the glow-worms light up their blue lamps to complete the revels. The delicate instrumentalist consists chiefly of a pair of large wings, thin and gleaming as strips of mica. Thanks to these dry sails, he fiddles away with an intensity capable of drowning the toad's fugue. His performance suggests, but with more brilliancy, more 'tremolo', the song of the common cricket.

These then are the principal participants in this musical evening. The scops owl, with his languorous solos; the toad, that tinkler of sonatas; the Italian cricket, who scrapes the first string of a violin, and the green grasshopper, who seems to beat a tiny steel triangle.

Left top: great green grasshopper (*Tettigonia viridissima*).
Left bottom: midwife toad (*Alytes obstetricans*) carrying eggs. (In fact it is the male of the species who carries the eggs. He carries them on his back legs for some weeks, then deposits them in water shortly before they are due to hatch.)

Right top: scops owl (*Otus scops*).
Bottom: Italian cricket (*Oecanthus pellucens*).

The Cricket

The cricket is extraordinary; of all the insects, he alone has a fixed home. Other insects may create marvels for their family — baskets of leaves, towers of cement, pits for ambush — but these are all temporary places that serve as nests or traps.

The cricket's home is a slanting gallery situated in the grass, on some sunny bank which soon dries after a shower. It is 23 centimetres long at most, as thick as one's finger.

The burrow is excavated from October onwards. The miner scrapes with his fore-legs and uses the pincers of his mandibles to extract the larger bits of gravel. I see him stamping with his powerful hind-legs, furnished with a double row of spikes; I see him raking the rubbish, sweeping it backwards and spreading it slantwise. Once the hole is a few centimetres deep, it is sufficient for the moment. From then on the work continues at a staggered pace, depending on the weather and the growth of the insect. Even in winter, if it is mild, it is not unusual to see the cricket shooting out rubbish, a sign of repairs and fresh excavations.

When April comes to an end, the cricket's song begins, at first in rare and shy solos, but soon developing into a general symphony in which each clod of earth boasts a performer. How is this wonderful music produced? Like all things of real value, it is very simple.

Fabre makes out that the structure of the sound mechanism is a simple one. Yet his description goes into such minute detail, as he examines the angles, veins, notches and wrinkles of each wing, that it is not easy to follow without having a live specimen at hand.

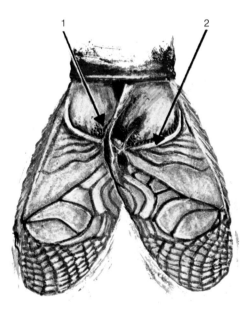

Below: view of cricket wings from above.
1 Scraper—the rough surface area on male wings.
2 File—this is on the underside of the male wings.

The field cricket produces sound by opening and closing its wings rapidly. With the closing of the wings, a hardened area of skin situated at both inner edges of the wings is rubbed over a series of minute teeth that form a miniature file. Movement of scraper across file causes the wings to vibrate at the high frequency of 5,000 cycles per second, producing a pure tone. Each closing action produces a single sound lasting only 25 milliseconds.

The commonest cricket song is the calling song, sung by the male to guide the female to his burrow. Once the male and female cricket have found each other a new song is sung: the courting song, which aids copulation, during which a minute sac of sperm the size of a pinhead is transferred from the male to the female's abdomen. In some species, after mating the males may sing a third type of song, known as the triumphal song. A fourth type of song is performed when rival male crickets invade each others' territory.

For the most part, Fabre's account of the cricket is a romantic view. He did, however, spend time rearing crickets in cages and watching the development of the young. It became evident to him that a large number of the offspring produced by one female is rapidly gobbled up in the wild by a host of predators, including ants and lizards, while later the survivors fall prey to the hunting wasp, the yellow-winged sphex. In Provence there are several species of cricket, but Fabre discounts the field cricket's relatives as wanderers of no fixed abode, and the only other one he describes in some detail is the Italian cricket, a slender, delicate insect that sings at night during the summer months.

37

The White-faced Decticus

The white-faced decticus heads the grasshopper clan in my district, both as a singer and as an insect of imposing presence. He has a grey costume, a pair of powerful mandibles and a broad ivory face. Though not plentiful, he does not let himself be sought in vain. In the height of summer we find him hopping in the long grass, especially at the foot of the sunny rocks where the turpentine tree takes root.

At the end of July I start a decticus menagerie. To make a vivarium, I stand a big wire gauze cover on a bed of sifted earth. The population numbers a dozen, and both sexes are equally represented.

The question of feeding perplexes me for some time. Judging by the locust, who consumes any green thing, it seems as though the regulation diet ought to be a vegetable one. I therefore offer my captives the tastiest and tenderest of green stuff: leaves of lettuce, chicory and corn salad. The decticus scarcely touch it; it is not the food for them. Perhaps something tough would suit their mandibles better. I try various grasses, which are accepted by the hungry ones; but it is not the leaves they devour, only the ears, whose still tender seeds they crunch with visible satisfaction. This taste for tender seeds surprises me: the name decticus means in Greek 'fond of biting'. Can this powerful jaw, of which I have to beware, possess no other function than to chew soft grains?

Now I find the real diet, the fundamental if not the exclusive one. Any fresh meat tasting of locust or grasshopper suits my ogres. As soon as the game is introduced into the cage, they stamp about and, hampered by their long shanks, dart forward clumsily. Speared by the hunter's fore-legs, the game is wounded in the neck. It is a very judicious first bite—the decticus seems to know exactly what it is doing. To overcome a prey that may escape so readily, it renders it helpless as quickly as possible, munching and extirpating the cervical ganglia, which is the main seat of innervation.

The white-faced decticus (Decticus albifrons) is one of the largest and most heavily built 'grasshoppers' to be found in Europe. It is actually an African insect, rarely found north of the Mediterranean region. With its 'impassive ivory face' it feeds mainly on other locusts and grasshoppers, especially the blue-winged locust. A related species of northern and central Europe is often called the wart-biter, because it was employed by Swedish peasants to chew off warts.

The dectitus has a fine song, consisting of a series of harsh metallic notes ('tick-tick') which build up to a crescendo and are then drowned by a droning bass. The song ends with a frantic rustle ('frrr-frrr-frrr').

Female decticus laying eggs, using her long ovipositor.

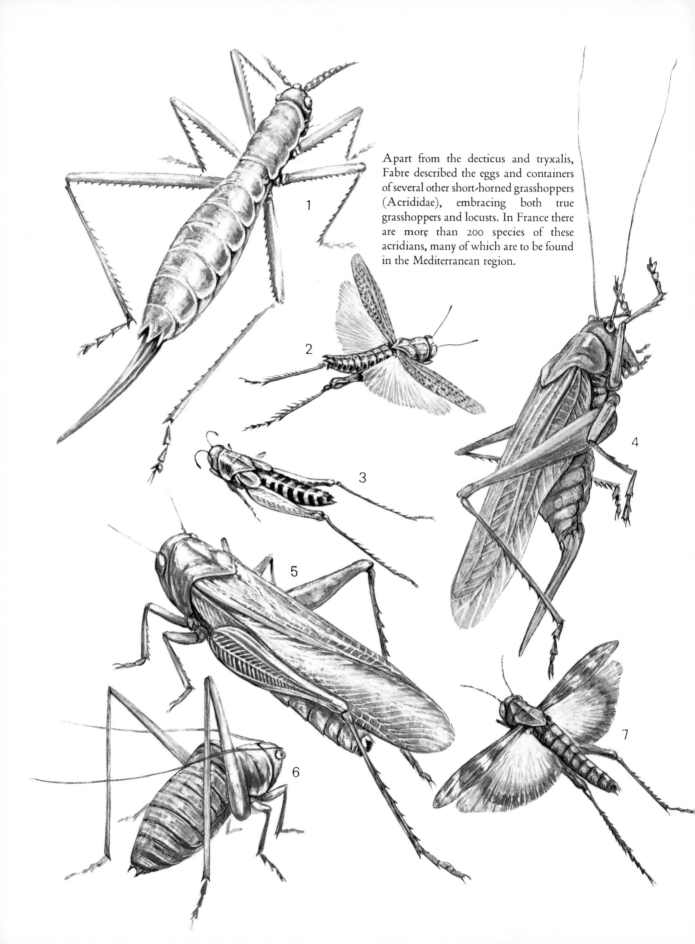

Apart from the decticus and tryxalis, Fabre described the eggs and containers of several other short-horned grasshoppers (Acrididae), embracing both true grasshoppers and locusts. In France there are more than 200 species of these acridians, many of which are to be found in the Mediterranean region.

Below: tryxalis or long-nosed cricket (*Acrida mediterranea*).

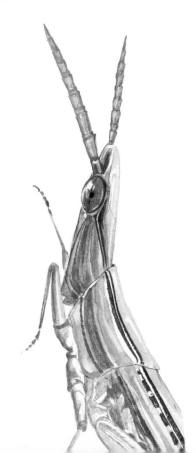

The Tryxalis and Other Grasshoppers

Though smaller than the grey locust, the tryxalis far exceeds her in slenderness and originality of shape. On our sun-drenched hillsides none has a leaping apparatus to compare with hers. What hind-legs, what extravagant thighs! They are longer than the creature's whole body. But with all this, the insect shuffles awkwardly along the edges of the vines, on the sand sparsely covered with grass; she seems embarrassed by her legs and with all this equipment the leap is disappointing, describing but a short curve reinforced by strong wings.

And then what a strange head!—an elongated cone shaped like a sugar loaf, whose point turns up in the air and has earned this insect the name of 'long nose'. At the top of the head are two large, gleaming oval eyes and two antennae, flat and pointed like dagger blades, which the tryxalis lowers with a sudden swoop to investigate any interesting object.

This is the creature that gives us most information about the method of egg-laying amongst the members of her tribe. In my cages, no doubt due to the boredom of captivity, it has never laid eggs on the ground. Instead, I have always seen it operating high up on the trellis of the cage, usually in October. Slowly, very slowly, it discharges its batch of eggs, which gush forth in a fine foamy stream, soon stiffening into a thick cord, knotty and oddly curved. This take nearly an hour. Then the thing falls to the ground, unnoticed by the mother.

This shapeless object is first straw-coloured, then darkens to a rusty brown. The first part is pure foam; the hind part alone is fertile and contains the eggs buried in a frothy matrix. These are amber-yellow, about a score in number and shaped like blunt spindles. By what mechanism does the tryxalis froth up her viscous product? She must certainly know the method of the preying mantis who with the aid of spoon-shaped valves whips and beats her foam into a fluffy omelette, but this grasshopper does all the work within the body. The glue is foamy from the moment it appears.

41

Locusts — their Function

If there is one peaceful and safe form of hunting, locust-hunting must be it. What delicious mornings we owe to it. What happy moments when the mulberries are black and my little helpmates can go pilfering here and there in the bushes. What memorable excursions on the slopes covered with sparse grass, tough and burnt yellow by the sun. I retain a vivid recollection of all this; and my children will do the same.

Little Paul has nimble legs, a ready hand and a piercing eye. He inspects the clumps of everlastings where the big grey locust solemnly nods his sugar-loaf head. He scrutinizes each bush from which this beast suddenly flies out like a little bird surprised by the hunter. Younger than her brother, Marie Pauline patiently watches for the Italian locust, with his pink wings and carmine hind-legs, but she really prefers another jumper. Her favourite wears a St Andrew's cross on the small of his back marked by four white, slanting stripes. With her hand raised in the air ready to sweep down, she approaches very softly, stooping low. Whoosh! That's done it. Quick, a screw of paper to receive the treasure which, thrust head first into the opening, plunges with one bound to the bottom of the funnel.

Thus are bags distended one by one. Before the heat becomes too great to bear, we are in possession of a number of varied specimens which, raised in captivity, will teach us something, if we know how to question them.

The first question that I put to my boarders is this: 'What function do you perform in the fields? You nibble the tops of the tough grasses which the sheep refuse to touch; you prefer the lean swards to the fat pastures. . .'

Eyed lizard (*Lacerta lepida*).

In September and October, the turkeys are driven into the stubble fields. The expanse over which the gobbling flock slowly spreads is bare, dry and burnt by the sun. What do the birds do in a desert like this, which simply reeks of famine? They cram themselves with locusts, which they snap up here and there, a delicious stuffing for their greedy crops.

When the guinea-fowl, that domesticated game-bird, roams around the farm, uttering her rasping note, what is it that she seeks? Seeds no doubt; but above all things locusts, which puff her out under the wings with a pad of fat and give greater flavour to her flesh.

If you are a sportsman and can appreciate the glory of the red-legged partridge, open the crop of the bird which you have just brought down. Nine times out of ten you will find it crammed with locusts. And as for the wheatear, who grows so disgracefully fat in September, the most significant item of his diet, in terms both of frequency and quantity, is the locust.

I have found locusts in the belly of that terror of the small girls of Provence, the 'rissado', or eyed lizard, who loves rocky shelters turned into a furnace by the torrid sun.

Without anything further being said about the devourers of this small game, it is easy to see the great usefulness of this animal, which by successive stages transmits to man, that most wasteful of eaters, the lean grass converted into exquisite fare. Gladly therefore would I agree with the Arab writer who said, 'It is very sure that, by the grace of God, grasshoppers were given to man for his nourishment.'

Wheatear (*Oenanthe oenanthe*) chasing locusts.

Praying mantis (*Mantis religiosa*) in spectral attitude. Fabre noted that the mantis only uses its imposing spectral attitude when confronting large prey such as the grey locust. For this shock tactic it employs a combination of eye spots, spread wings and puffing sounds from the abdomen. This performance stops any insect in its tracks just long enough for the mantis to pounce.

The Praying Mantis

Her Hunting

Folk hereabouts call her 'lou Prego-Dieu', the animal that prays to God. She is the tigress of the peaceful insect tribes, the ogress who waits in ambush. Apart from her lethal implement, the mantis has nothing to inspire dread. She is not without a certain beauty in fact, with her slender figure, her elegant bust, her pale green colouring, and her long gauze wings.

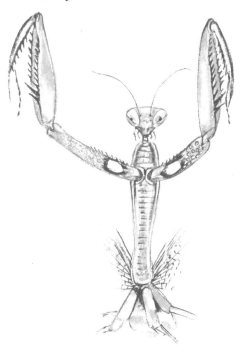

43

Praying mantis hunting, showing movement of the fore-limbs as it strikes.

Her murderous mechanisms are her fore-legs. The haunch is uncommonly long and powerful; its function is to throw forward the rat trap, which does not await its victim but goes in search of it. The base of the haunch is adorned on the inner surface with a pretty black mark with a white spot in the middle. The thigh, longer still, carries on the front half of its lower surface two rows of sharp spikes. It is, in short, a saw with two parallel blades, separated by a groove in which the legs lie when folded back. The leg is likewise a double-edged saw, with very many smaller teeth. It ends in a strong hook which has a double blade like a curved pruning knife. Locusts of all kinds, butterflies, dragonflies, large flies and bees are what we usually find in the lethal limbs. In my cages, the daring huntress recoils before nothing. The facts are worth describing.

At the sight of the quarry, the mantis gives a convulsive shiver and suddenly adopts a terrifying posture. The wings, spread to their full extent, stand erect like parallel sails towering over the back; the tip of the abdomen, curling upwards, rises and falls, relaxing with short jerks. It reminds one of the puffing of a startled adder.

Planted squarely on its four legs, the insect holds itself upright, its legs open wide to form a cross with the body and reveal the black spots, faint imitations of the peacock's tail. Motionless in this strange posture, the mantis watches its prey, which remains rooted to the spot, face to face with death.

Then the two grapnels fall, the claws strike, the double saws

close and clutch. In vain the poor wretch protests; he chews space with his mandibles and, kicking desperately, strikes nothing but the air. The mantis furls her wings, resumes her normal posture; and the meal begins.

Praying mantis laying eggs.

Her Love-Making

It is near the end of August. The slender male makes eyes at his strapping companion; he turns his head in her direction; he bends his neck and throws out his chest. Motionless for a long time, he contemplates the object of his desire. She does not stir, as though indifferent. The lover catches a sign of acceptance, a sign whose secret I do not know. He goes nearer; then suddenly spreads his wings, which quiver with a compulsive tremor. He rushes, small as he is, upon her back, clings on as best he can, steadies his hold. At last coupling takes place, a long drawn out affair, sometimes lasting for five or six hours.

They end by separating, but only to unite again in a less romantic fashion. If the poor fellow is loved by his lady as a lover, he is also loved as a piece of highly flavoured game; and that same day he is seized by her. She begins by gnawing his neck, then eats him deliberately by little mouthfuls, leaving only the wings.

I have seen one and the same mantis use up to seven males. She takes them to her bosom and makes them pay for the nuptual ecstasy with their lives.

A transverse section through the ootheca or egg case, showing the elongated eggs in separate compartments. The young nymphs hatch out from pores in the top of the ootheca.

45

The Empusa

The larva of the empusa is certainly the strangest creature among the insects of Provence: a slim, swaying thing of so fantastic an appearance that uninitiated fingers dare not lay hold of it. The children of my neighbourhood, impressed by its startling shape, call it 'the Devilkin'. One comes across it, though always sparsely, in spring, up to May, in autumn, and sometimes in winter if the sun is strong. The tough grasses of the wastelands, the stunted bushes which catch the sun and are sheltered from the wind by a few heaps of stones are the chilly empusa's favourite abode.

What a queerly shaped head it has! A pointed face, with walrus moustaches, large goggle eyes and, on the forehead, a mad, unheard of thing—a sort of tall mitre, an extravagant head-dress that juts forward, spreading right and left into peaked wings and cleft along the top. What does the Devilkin want with that monstrous pointed cap, the like of which no wise man of the East, no astrologer of old ever wore? The dress is commonplace; grey tints predominate. Towards the end of its larval life, it begins to offer a glimpse of the adult's richer form and becomes striped, still very faintly, with pale green, white and pink. Already the two sexes are distinguished by their antennae. Those of future mothers are thread-like; those of the males are distended into a spindle at the lower half, forming a case or sheath from which graceful plumes will spring at a later date.

The transformation into adults comes in May. Of its youthful eccentricities the adult retains the pointed mitre, saw-like arm guards, knee-pieces, scales on the belly; but the abdomen is now no longer twisted as it was into a crook, and the animal is comelier to look upon.

Big eaters are naturally quarrelsome. The mantis, bloated with locusts, soon becomes irritated and shows fight. The empusa with her frugal meals does not indulge in hostile demonstrations. Unknown also are tragic courtships. The male is enterprising and is subjected to a long trial before succeeding with his mate. The feathered groom then retires and does his

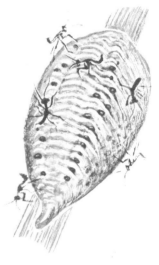

Above: mantis ootheca.

Below: ootheca of the empusa (*Empusa pennata*). 'Those exit holes, so regular in shape and arrangement, present the appearance of two dainty mouthorgans placed in juxtaposition. Each of them corresponds with a cell containing two eggs. The eggs in all, therefore, amount to about a couple of dozen.'

little bit of hunting, with no danger of being gobbled up. The two sexes live together in peace and mutual indifference until the middle of July. Then the male, grown old and decrepit, hunts no more, becomes shaky in his walk and at last collapses on the ground.

The laying follows close upon the disappearance of the males. The empusa, when about to build her nest, has not the round belly of the mantis rendered heavy and inactive by her fertility. Her nest, fixed upon a straw, a twig, a chip of stone, is quite as small as that of the dwarf mantis and measures no more than a centimetre at most in length. The general shape is that of a trapezoid, of which the shorter sides are sloping and slightly convex. A very thin grey-wash, formed of dried foam, covers the empusa's work, while six or seven hardly perceptible furrows divide the sides into curved sections. After the hatching, a dozen round orifices open on the top of the building, in two alternate rows. It is unfortunate that I have never seen this hatching taking place.

While all other mantids spend the winter in the egg stage, the larvae of the empusa emerge in the autumn and the adults can be seen hunting in warm places as early as May.

In Fabre's day the ootheca was used by the country people of his district as a remedy for chilblains. However, Fabre tried it on himself and his family to no avail and so deduced that the only connection between chilblains and the mantis's ootheca was the word 'tigno', the Provençal name for both.

Below left: 'Devilkin', the nymph of the empusa. While all other mantids of the region overwinter in the egg stage, the nymph of the empusa emerges in the autumn, spends the winter in this stage and transforms into an adult late in the spring.
Below right: adult empusa.

Hunting Wasps

One winter evening, when the rest of the household was asleep, I sat reading beside the still-warm ashes of a stove. My book made me forget for a while the cares of tomorrow, those heavy cares of a poor professor of physics. It was a monograph by the then father of entomology, Léon Dufour, on the habits of a beetle-hunting wasp.

Needless to say, my interest in insects dates from long before then: from my earliest childhood I have delighted in beetles, bees and butterflies; as far back as I can remember I have been enraptured by the splendour of a ground beetle's wing cases or a swallowtail's wings. The fire was laid; the spark to kindle it was absent. Léon Dufour's essay provided that spark.

New lights burst forth: I received a sort of mental revelation. There was more to science than arranging pretty beetles in a cork box and giving them names and classifying them; there was something much finer: a close and loving study of insect life. The examination of the structure and especially the faculties of each species.

When young, the hunting wasps feed contentedly on the nectar of flowers, but later in the season they turn their attention to live prey, fresh meat for their developing young, who are kept safe in underground burrows excavated by the female of the species.

1 *Ammophila*
2 *Bembex*
3 *Cerceris*
4 *Pompilus*

Ammophilae — Caterpillar Hunters

A slender waist, a slim shape; an abdomen tapering very much at the upper part and fastened to the body as though by a thread; black raiment with a red sash across the belly: there you have a summary description of these burrowers.

If the name were not so pleasing to the ear, I would readily quarrel with the term *Ammophila*, which means 'sandlover', as being too exclusive and often erroneous. The real lovers of sand, of dry, dusty, streaming sand, are the *Bembex*, who prey on flies, but the caterpillar hunters have no predilection for ordinary shifting sand, and even avoid it as being liable to landslips at the slightest provocation. Their perpendicular shaft, which has to remain open until the cell receives the provisions and the egg, requires a firmer setting if it is to avoid the risk of being prematurely blocked. What they want is a light soil, easily tunnelled, in which the sandy element is cemented with a little clay and lime. Edges of paths, sunny banks where the grass is rather bare — those are their favourite spots. In spring, quite early in April, we see the hairy *Ammophila* there; when September and October come, we find the sandy *Ammophila*, the silvery *Ammophila* and the silky *Ammophila*.

Above: *Ammophila* carrying paralysed caterpillar.
Right: *Ammophila* dragging caterpillar into burrow.
Far right: caterpillar in cell with single egg laid.

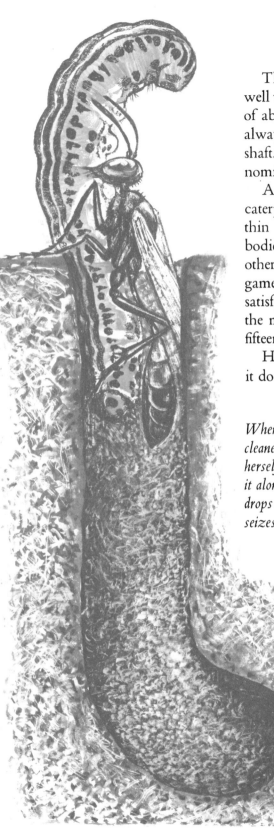

The burrows of all four consist of a vertical shaft, a sort of well with the diameter of at most a thick goose quill and a depth of about five centimetres. At the bottom is the cell, which is always solitary and consists of a mere widening of the entrance shaft. It is, when all is said, a poor lodging, obtained economically by one day's work.

As to their game, all these wasps hunt for their larvae the caterpillars of moths. The silky *Ammophila* selects those long thin caterpillars that walk by looping and unlooping their bodies. Five of these are used to stock up one cell. The three other *Ammophilae* give only one caterpillar to each larva, the game in their case being big and plump, capable of amply satisfying the grub's appetite. For example, I have taken from the mandibles of the sandy *Ammophila* a caterpillar weighing fifteen times as much as the wasp herself.

How do these wasps paralyse such elongated prey, possessing as it does twelve distinct nerve centres apart from its feeble brain?

When the Ammophila *has stung and paralysed her caterpillar and cleaned herself after the struggle, she turns the body on its back, places herself astride the prey and, grasping its throat with her mandibles, hauls it along. It is a laborious business. Eventually, getting nearer home, she drops her load and goes off to open the shaft. She then backs into the hole, seizes the caterpillar by the head and pulls it down the shaft into her cell.*

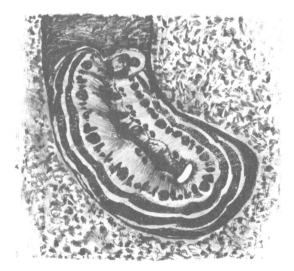

At the beginning of my investigations I twice watched the silky *Ammophila* in action and as far as I could see from so swift an operation, the wasp's sting was applied once and for all to the victim's fifth or sixth segment, these being legless. In putting these two nerve points out of action the wasp had deftly separated the two main locomotory centres on either side, rendering the caterpillar motionless. It is on these crucial segments that the egg is laid; they offer the newly hatched grub a safe food store on which to start its life.

But the sandy *Ammophila* and the hairy *Ammophila*, the latter in particular, capture enormous victims. Will this giant prey be treated in the same manner as the frail loopers? I have been fortunate enough to see the hairy *Ammophila* use her scalpel and never has the science of instinct shown me anything more exciting. There she was busily scratching the ground at the foot of a tuft of thyme. She ran hurriedly this way and that inspecting every crevice that would give access to what lay below. She was not digging herself a home, but hunting — like a dog trying to dig a rabbit out of his hole. Then, roused by what was going on above, a big grey caterpillar surfaced. That settled him; the huntress was on the spot at once, gripping him by the skin of his neck and holding him tight. Perched on this monster's back, the wasp bent her abdomen and deliberately, like a surgeon thoroughly acquainted with her patient's anatomy, drove her lancet into the belly of each of the victim's segments from first to last. Not a ring was left without a stab.

The wasp acts with a precision that would make science turn green with envy; she knows her victim's complex nervous system and reserves each dagger thrust for the successive nerve ganglia of her caterpillar. I say she knows; what I should say is, she behaves as though she knew. Her act is simple inspiration. Animals obey their compelling instinct, without realizing what they do.

Sandy *Ammophila* (*Ammophila sabulosa*).

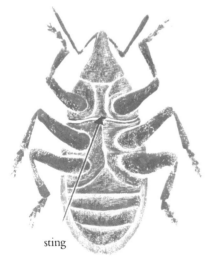

Top: the wasp *Cerceris tuberculata* stinging the weevil *Cleonus ophthalmicus*. Weight of wasp 1.5 g; weight of prey 2.5 g.

Above: she stings the prey on its underside between the first and second pairs of legs. At this vital point she puts out of action the thoracic ganglia which control all movement of the beetle's legs and wings.

sting

Cerceris — the Sting

This wasp establishes her home in the steep roadside banks, and in the sides of the ravines hollowed by the rains in the sandstone. These conditions are common in the neighbourhood of Carpentras and it is here that I have observed *Cerceris tuberculata* in the largest numbers.

Yet she does not rest content with the choice of this vertical site; she takes other precautions to guard against the inevitable rains of the season. If there is some bit of hard sandstone projecting like a ledge, she will contrive her gallery under this shelter.

Though no sort of communism exists among them, these insects nevertheless like to associate in small numbers; and I have always observed their nests in groups of about ten at least, sometimes close enough to touch each other.

The victim on which the *Cerceris* chooses to feed her grubs is a large-sized weevil. We see the kidnapper arrive heavily laden, carrying her victim between her legs, body to body, head to head; then plump down at some distance from her hole to complete the rest of the journey without the aid of wings.

54

I have seen the wasp confronting her victim. Gripping its snout with her powerful mandibles, she soon had it at her mercy. While the weevil reared up on its six legs, the wasp pressed her fore-feet violently on his back, as if to force open some ventral joint. Next I saw the assassin's abdomen slip under the weevil's belly, bend into a curve, and dart its poisoned lancet swiftly, two or three times into the joint between the first and second pair of legs. The victim fell motionless for all time, as though struck by lightning. It was terribly and at the same time wonderfully quick.

Thus these powerfully built weevils, which, if pierced with a pin and fixed on an insect collector's fatal sheet of cork, struggle for days and weeks, here instantly lose all power of movement from the effect of a tiny prick which inoculates them with an invisible drop of venom. Chemistry has no poison so potent in so minute a dose. So we turn to physiology and anatomy rather than to toxicology in order to grasp the cause of this instantaneous annihilation. What is there, then, at the point where the sting enters?

Fabre as mischief-maker and curious observer robbed Cerceris *of her prey several times just as she had deposited it at the base of the sandstone slope before hauling it up to her burrow. Unperturbed by her loss, the wasp resumed hunting almost immediately and in less than ten minutes returned with fresh prey. In order to see how the sting was administered, Fabre first scoured the neighbourhood for the particular weevil. 'Two days' search produced three weevils, flayed, covered with dust, minus antennae or legs, maimed veterans!' Offering live weevils at the burrow entrance was a failure: the wasp ignored this potential prey. So, instead, Fabre removed the paralysed prey and substituted fresh prey just as the wasp unloaded at the bottom of the slope. Then he could watch* Cerceris, *the 'scientific slaughterer' in action. By comparing the nervous system of various beetles, he found that the only beetles of convenient size in which the nervous system was so arranged that they could be immobilized quickly and efficiently were the weevils. The scarabs and scolytes have a similar nervous system. Fabre concluded that it must be the size of these that make them unsuitable prey for this particular hunting wasp. Simulating the wasp's sting by using a fine-nibbed pen tipped with a drop of ammonia, Fabre produced the same effect—complete instant paralysis.*

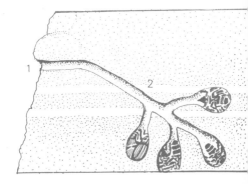

Above:
1 A sandstone slope with overhanging ledge. This is where *Cerceris* drags her selected weevil up the bank to the burrow entrance.
2 The burrow (45 cm long) divides into cells, usually stocked with five or six weevils.

Below:
3 An enlarged section showing position of laid eggs. As with most of the hunting wasp group, the egg is usually positioned where the sting is made.

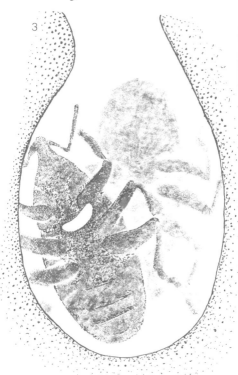

The Fly Hunt

One of my favourite spots for the observations which I am about to describe is not far from Avignon on the right bank of the Rhône opposite the mouth of the river Durance. Issarts wood is a coppice of holm oaks no higher than one's head and sparingly distributed.

On certain afternoons in the dog-days of July and August I used the shelter of a large umbrella. I was not the only one to profit by its shade; I was generally surrounded by numerous companions. Gadflies of various species would take refuge under the silken dome and sit peacefully on every part of the tightly stretched cover. I was rarely without their society when the heat became overpowering. To while away the hours when I had nothing to do, it amused me to watch their great golden eyes, which shone like carbuncles under my canopy.

One day, bang! The tight cover resounded like the skin of a drum. Perhaps an oak had dropped an acorn on the umbrella. Presently one after the other, bang, bang, bang! I left my tent and inspected the neighbourhood—nothing! The same sharp sound came again. I looked up and the mystery was explained. The *Bembex* of the neighbourhood were impudently penetrating my shelter to seize flies on the ceiling. I had only to sit still and look. Every moment a *Bembex* would enter, swift as lightning, and dart up to the silken ceiling, which would resound with a sharp thud. The struggle did not last long, and the wasp would soon retire with a victim between her legs. At this sudden irruption, which was slaughtering them one after the other, the dull herd of gadflies drew back a little all round, but did not quit the treacherous shelter. It was so hot outside! Why get excited?

Below: *Bembix rostrata* carrying blue-bottle (*Calliphora vomitoria*).

56

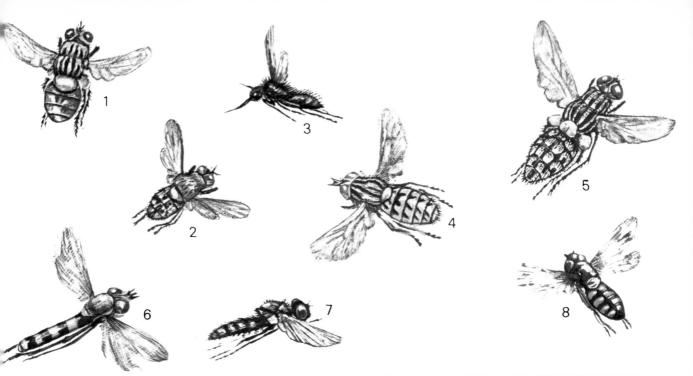

Typical flies hunted by *Bembix rostrata*:
1 Stable fly (*Stomoxys calcitrans*)
2 Cluster fly (*Pollenia rudis*)
3 Bee fly (*Geron gibbosis*)
4 Horse fly (*Tabanus bromius*)
5 Flesh fly (*Sarcophaga carnaria*)
6 Hover fly (*Syrphus corollae*)
7 Hover fly (*Sphaerophoria scripta*)
8 Hover fly (*Pipiza nigripes*)

Like a mother bird caring for her young, the female Bembex (Bembix) provides her grub with a constant supply of fresh food. Her prey consists of flies. Any type will do, provided the size is right. First she chooses a tiny morsel on which to lay the egg. Then, as her larva's appetite increases, she forages further afield for larger and larger game.

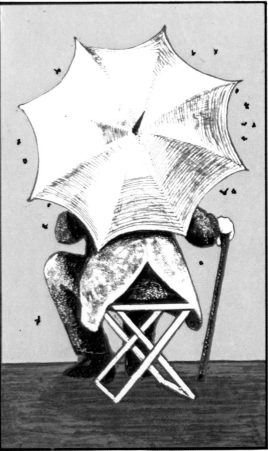

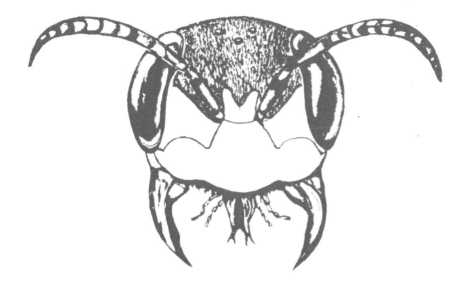

Philanthus — the Bee Killer

Philanthus kills its bee instead of paralysing it. Once having committed the murder, it does not release the bee for a moment, but holding it tightly clasped with its six legs pressed against its body, it commences to ravage the corpse. I see it squeezing the bee's stomach, compressing it with its own abdomen, crushing it as if in a vice.

These various manipulations, above all the compression of the throat, have the desired effect: the honey in the stomach of the bee ascends to the mouth. I see the drops of honey welling out, lapped up by the glutton as soon as they appear. *Philanthus* takes this odious meal lying on its side with the bee between its legs. The meal lasts often half an hour or longer.

I am far from denying that *Philanthus* has honest methods of earning a living; I see it among the flowers, no less busy than others of its tribe, peacefully drinking their cups of nectar. Indeed, the male, being stingless, knows no other means of supporting himself. But the mothers, though not neglecting the flowers, live by piracy as well.

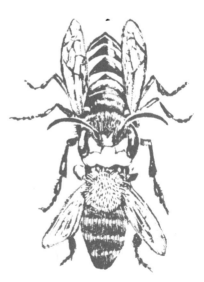

58

Bees intended as food for the larvae are stung under the chin like the others; they are true corpses, manipulated, squeezed and exhausted of their honey just like the others. Why are the bees robbed of their honey before they are given to the larvae? It occurs to me that beefsteak garnished with 'confiture' is not to everyone's taste, and the bee sweetened with honey may well be distasteful or even harmful to the larvae. I smear a dead bee with honey, lightly varnishing it by means of a camel's hair brush. The grub bites on the honeyed bee, draws back as though disgusted, hesitates for a long time, then urged on by hunger begins again; but finally it refuses to touch the bee. For a few days it pines upon its rations, which remain almost intact, then dies. By emptying the stomach of the bee, the mother is performing the most praiseworthy of duties: she is guarding the family against poison.

Above: bee fly (*Bombylius*), occasionally mistaken for a bee by *Philanthus*. She quickly realizes her mistake and immediately disposes of the dead fly.

Below: three worker honey bees.

Sphex

As far as I know, the French fauna has only three species of *Sphex*, all lovers of the hot sun. These are the yellow-winged *Sphex*, the white-edged *Sphex*, and the Languedocian *Sphex*. To feed their grubs, all choose orthopterans; the first hunts crickets, the second locusts and the third ephippigers.

Yellow-winged *Sphex* (*Sphex flavipennis*) with prey—field cricket (*Gryllus capestris*).

Many a time I have surprised the Languedocian *Sphex* sunning herself on a vine leaf. The insect, spread out flat, basks voluptuously in the heat and light. It may be that this couch also serves as an observatory from where she scans the surrounding countryside for prey. Her exclusive prey is the *Ephippiger* of the vine, of which she selects only females whose bellies are swollen with a mighty cluster of eggs.

The bulk and weight of the prey entirely reverses the usual order followed by other hunting wasps; that is, first digging the burrow and then stocking it with prey. The Languedocian *Sphex* first paralyses the prey; then excavates a burrow as near as possible to the spot where the victim lies. As each victim is caught, a fresh excavation is made, a fresh burrow with but a single chamber.

White-edged *Sphex* (*Prionyx kirbyi*) with prey—blue-winged locust (*Oedipoda coerulescens*).

Two typical experiments carried out by Fabre, demonstrating the Sphex's inability to improvise:
1. *The Languedocian* Sphex *drags its prey, the* Ephippiger, *along by the antennae. Fabre cut these antennae off, and found that the wasp merely changed its grip and held its prey lower down at the stumps. When these were cut off, the wasp still held it by the mouthparts. When all mouthparts were removed, the wasp abandoned the prey altogether, even though it could have gripped it by the legs or abdomen.*

2. *Again using the Languedocian* Sphex, *Fabre opened up the burrow and removed the prey just as the wasp was sealing up the burrow. Although the wasp went repeatedly in and out of the burrow, she soon resumed sealing up the entrance, and never returned with fresh prey. Fabre was convinced that the* Sphex's *behaviour was completely instinctive. 'I conclude therefore: instinct knows everything in the undeviating paths marked out for it; but it knows nothing outside those paths.'*

Languedocian *Sphex* (*Palmodes occitanicus*) with prey—bush cricket (*Ephippiger ephippiger*).

60

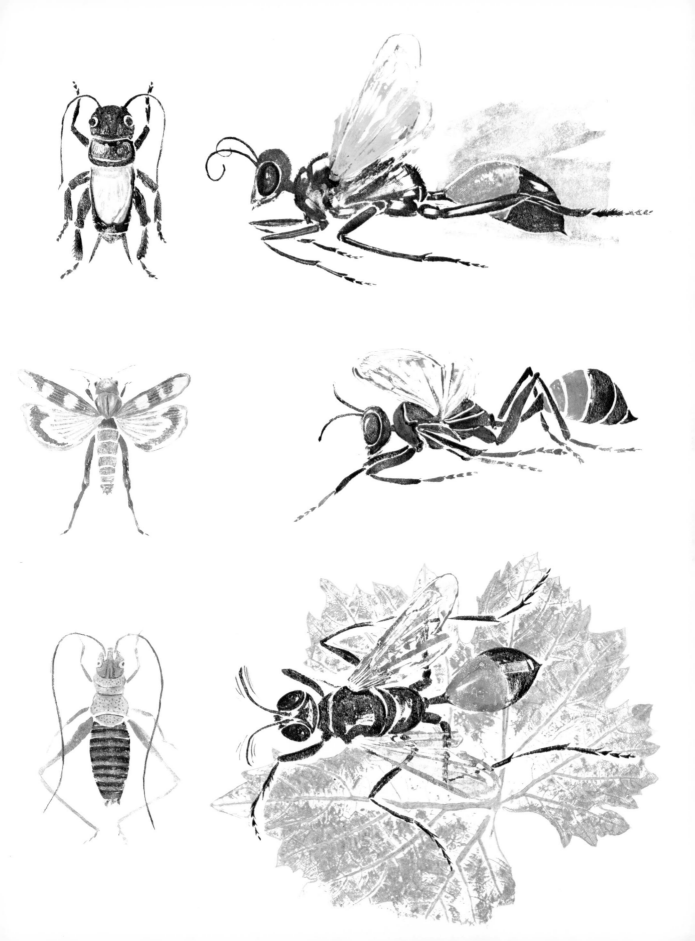

Insects as Architects

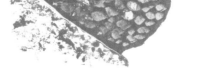

The industry of insects, especially that of the bees and wasps abounds in tiny marvels.

The *Chalcidoma* of the walls, when building on a pebble, first raises a turret of faultless geometrical proportions. The dust scraped from the hardest spots in the highways kneaded with saliva provides the mortar. To make a more solid job, and also to economize on cement, tiny bits of gravel are encrusted in the outer surface before the material sets. In this way the initial building becomes a rustic rockwork fortress, which is quite pretty to look at.

The simplest of round bodies, the cylinder, is the model for the jam pot wherein the *Pelopaeus* stacks her spiders. With mud collected from the edge of a pool, the huntress begins by building a turret ornamented with diagonal lozenges. Unhampered by its surroundings, this structure, the first of a group, is fashioned like the segment of a twisted column. But other cells are to follow; leaning one against the other, these produce a mutual distortion. A thick layer of cement ends by deforming the structure altogether.

Nest of mason bee (*Chalcidoma muraria*).

The pottery of *Eumenes* is of a high order: it favours a bulging cupola, like that of a Turkish mosque or the Moscow basilica. At the summit of the dome is a short opening, through which the caterpillars intended for the larva's consumption are introduced. When the larder is full and the egg slung from the ceiling by a thread, the bell-mouthed neck is closed with a clay stopper. Yet this elegant structure is doomed to disappear. Once other cells have been added, there remains barely any evidence of the work of an expert dome builder in the shapeless blob which is the finished task.

Nest of potter wasp (*Sceliphron*).

Nest of potter wasp (*Eumenes pomiformis*).

If an insect were to build a general shelter, in which each larva had its individual box, what would this building be? If I were to draw you the prettiest child's balloon ever inflated in toyland, it would be exactly like the nest of the *median* wasp (*Vespa media*). The person who gave me this marvel found it hanging from the lower edge of a shutter that was left open for the greater part of the year. With paper manufactured by herself, tough and flexible as the silk papers of China and Japan, the wasp had contrived to expand her work into a segment of an ellipsoid, with a cone added by means of a gentle curve. A similar association of forms artistically combined is found in the sacred beetle's pears.

Ill-defined spiral bands tell us how the wasp went to work. With a pellet of paper pulp in her mandibles, she moved downwards in a slanting direction following the margin of the part already constructed and leaving a ribbon of her material still quite soft and impregnated with saliva. The work was discontinued and resumed hundreds and hundreds of times. There was even collaboration between several builders. The foundress of the city, the mother, alone at the outset was able only to make a rough beginning of the roof; but offspring arrived, the workers, charged with continuing and enlarging the dwelling in order to provide the one mother with a lodging to contain all her eggs. This gang of paper makers, coming one by one to take part in the labour, achieved perfect regularity. By gradual degrees the spacious dome of the summit decreases in diameter; by degrees it tapers into a cone and ends in a graceful neck. Individual and almost independent efforts result in a harmonious whole.

Right: nest of a social wasp (*Vespa media*) showing entrance at bottom. Above: section of outer bark pulp casing removed, showing internal cellular structure.

The Crab Spider

Crab spiders (*Thomisus onustus*) camouflaged on a *Cistus*. Note honey bee ambushed by spider.

Thomisus is passionately addicted to the pursuit of the domestic bee. This murderess of the bee is of a chilly constitution; in our region, she hardly ever moves away from the olive districts. Her favourite shrub is the white-leaved rock-rose with the large, pink, crumpled, ephemeral blooms that last a morning and are replaced next day by fresh flowers that have blossomed in the cool dawn.

Here the bees plunder enthusiastically, fussing and bustling in the spacious whorls of the stamens. The spider posts herself in her watch house under the rosy screen of a petal. Cast your eyes over the flower. If you see a bee lying lifeless with legs and tongue outstretched, draw nearer; nine times out of ten the *Thomisus* will be there. The thug has struck her blow; she is draining the blood of the departed.

This cutter of bees' throats is a pretty creature despite an unwieldy paunch fashioned like a squat pyramid and embossed on the base of either side with a pimple shaped like a camel's hump. The skin, more pleasing to the eye than any satin, is milk-white in some, in others lemon-yellow. Some are fine ladies who adorn their legs with a number of pink bracelets or carmine arabesques. Novice fingers that shrink from touching any other spider allow themselves to be enticed by these attractions.

This gem among spiders makes a nest worthy of its architect. A lover of high places, the *Thomisus* selects as the site of her nest one of the upper twigs of the rock-rose, a twig withered by the heat, and possessing a few dead leaves that curl into a little cottage.

Ascending and descending with a gentle swing, the living shuttle, swollen with silk, weaves a bag whose outer casing becomes one with the dry leaves around. The nest is more than a place of rest after the fatigues of her confinement: it is a guard room, an inspection-post where the mother remains sprawling until the youngsters' exodus.

65

The Clotho

The *Clotho* is above all a talented spinstress. We find her on the rocky slopes in the scorched and sun-blistered land of the olive. Turn over the flat stones, those of a fair size; search, above all, the piles which the shepherds set up for a seat to watch over the sheep. Do not be too easily disheartened: the *Clotho* is rare. If we are lucky, we shall see, clinging to the lower surface of the stone, a structure shaped like an overturned cupola and about half the size of a tangerine. The outside is encrusted with small shells, particles of earth and, especially, dried insects. The edge is scalloped into a dozen angular lobes the points of which spread and are fixed to the stone. In between these straps is the same number of spacious inverted arches. The whole resembles the Ishmaelite's camel-hair tent, but upside down. A flat roof, stretched between the straps, closes the top of the dwelling.

Inside, the *Clotho* is quite fastidious. Her couch is more delicate than swan's down. This exquisite retreat demands perfect stability, especially on gusty days when sharp draughts penetrate beneath the stones. Take a careful look at the web. The arches that gird the roof with a balustrade and bear the weight of the web are fixed to the slab by their extremities. In addition, from each point of contact there issues a cluster of diverging threads that creep along the stone. I have measured some fully 23 centimetres long. These are the cables; they represent the ropes and pegs that hold the Arab's tent in position.

The hanging shells are I suspect used for ballast and balance. The house spider, who spins her web in the corner of a wall, prevents the web from losing its shape by loading it with crumbling plaster. So the *Clotho* in a more ornamental way uses the empty and the occupied shells of the smaller Mollusca.

The *Clotho*, who is not only nocturnal but also excessively shy, conceals her habits from us; she shows us her works but hides her actions, especially that of egg-laying, which I estimate takes place in October. The mother sits upon her heap of eggs with the same devotion as a brooding hen. The hatching takes place early; November has not yet arrived when the young,

Purse-shaped web of the *Clotho* spider.

After venturing out among the rocks under which she builds her web, the *Clotho* (*Uroctea durandi*) digests the small snails, throwing them over the side and leaving them dangling by remnants of web.

miniature replicas of the adult, hatch out. Packed close together, they spend the whole wintry season in the tent with mother watching over them. When the summer heat arrives in June, the young ones, probably aided by their mother, pierce the walls, leave the tent, take the air on the threshold for a few hours, then fly away, carried to some distance by a funicular aeroplane, the first product of their spinning mill.

Narbonne Lycosa — the Black-bellied Tarantula

My district does not boast the ordinary tarantula, but it possesses an equivalent in the shape of the black-bellied tarantula, or Narbonne *Lycosa*, half the size of the other, clad in black velvet with brown chevrons on the abdomen and grey and white rings on the legs. Her home is the dry pebbly ground covered with sun-scorched thyme.

The dwelling is a pit about 23 centimetres deep, perpendicular at first, and then bent elbow-wise. The average diameter is $2\frac{1}{2}$ centimetres. On the edge of the hole stands a kerb formed of straw, bits and scraps of all sorts, and even small pebbles the size of a hazelnut. The whole is kept in place and cemented with silk.

From her little turret, the *Lycosa* lies in wait for the passing locust. She gives a bound, pursues the prey and

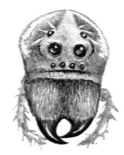

Top: looking down the burrow of the Narbonne *Lycosa* (*Lycosa narbonnensis*). Above: frontal view of head, showing the size and power of the tarantula's 'jaws'.

Right: vertical section of the burrow, showing turret.

suddenly deprives it of motion with a bite in the neck. The game is consumed on the spot, or else in the lair. This sturdy huntress is not a drinker of blood; she needs solid food, food that crackles between the jaws. She is like a dog devouring a bone.

Would you care to bring her to the light? Insert a straw into the burrow and move it about. Uneasy as to what is happening above, the recluse hastens to climb up and stops, in a threatening attitude, at some distance from the orifice. You see her eight eyes gleaming in the dark; you see her powerful poison fangs yawning, ready to bite. Br-r-r! Let us leave the beast alone.

The *Lycosa* —
The Family

At the beginning of the month of August, the children call me to the far side of the enclosure, rejoicing in a find which they have made under the rosemary bushes. It is a magnificent *Lycosa* with an enormous belly, the sign of an impending delivery. The obese spider is gravely devouring something in the midst of a circle of onlookers. What? The remains of a *Lycosa* a little smaller than herself—the remains of her male. It is the end of the tragedy that concludes the nuptuals. The sweetheart is eating her lover.

Early one morning ten days later, I find her preparing for her confinement. First, a silk network is spun on the ground. It is coarse and shapeless, but firmly fixed. On this foundation, which acts as a protection from the sand, the *Lycosa* fashions a round mat, the size of a two-franc piece and made of superb white silk. This silk disk is increased in thickness. The piece thus becomes bowl-shaped, surrounded by a wide, flat edge.

The time for laying has come. With one quick emission, the viscous pale-yellow eggs are laid in the basin. The spinnerets are once more set going. With short movements, as the tip of the abdomen rises and falls, they cover up the exposed hemisphere.

At the end of the summer Fabre noticed how the female *Lycosa* came up to the turret of her burrow and, holding the egg basket with her hind-legs, presented it to the sun, gently revolving it until each part had been warmed. The female would emerge in the morning (provided it was warm) and spend half the day patiently holding out her 'white pill bulging with germs'. This activity lasts from three to four weeks.

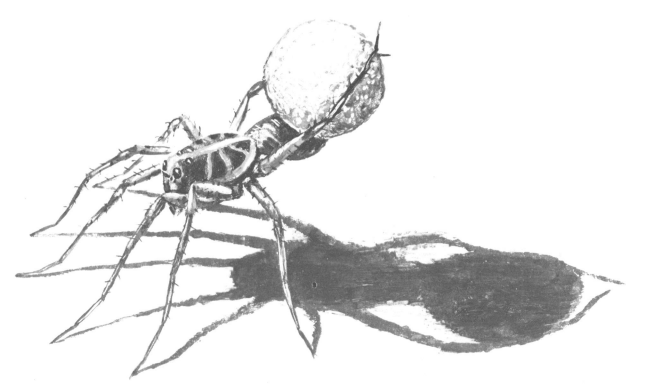

The work of spinning is continued for a whole morning. Now, worn out with fatigue, the mother embraces the white-silk cherry-sized cocoon and remains motionless. I shall see no more today.

Next morning, I find the spider carrying a bag of eggs slung from her stern. From that time until the hatching, she does not leave go of the precious burden which, fastened to the spinnerets by a short ligament, drags and bumps along the ground.

Formerly the *Lycosa* came out in the sun for her own sake. Leaning on the parapet of her burrow, she had the front half of her body outside the pit and the hinder half inside. When carrying the egg-bag, the spider reverses her posture: the front is in the pit, the rear outside. Gently she turns, so as to present every side to the life-giving rays. This goes on for half a day, so long as the temperature is high; and it is repeated daily, with exquisite patience, for three or four weeks.

In the early days of September, the young ones, who have been some time hatched, are ready to come out. The whole family emerges from the bag straight away. Then and there the youngsters climb onto the mother's back. Huddled together, sometimes in two or three layers, the little ones cover the whole back of the mother, who for seven or eight months to come will carry her family night and day. Nowhere can we hope to see a more edifying picture than that of the *Lycosa* clothed in her young.

The young emerge in September and straight away hop on to their mother's back, where they remain for seven months without feeding. It is not until the following March that the youngsters leave their mother in small batches to hunt for food and excavate their own burrows.

71

The Banded Epeira

In bearing and colouring *Epeira fasciata* is the handsomest of the spiders of the south. On her fat belly, a mighty silk warehouse as large as a hazelnut, are alternate yellow, black and silver sashes. Around the portly abdomen, the eight long legs with their dark and pale brown rings radiate like spokes.

Her hunting weapon is a large, upright web, whose outer boundary is fastened to the neighbouring branches by a number of moorings. The structure is that adopted by the other weaving spiders. Straight threads radiate at equal intervals from a central point. Over this framework runs a continuous spiral thread, forming chords or crossbars from the centre to the circumference. It is magnificently large and magnificently symmetrical.

Starting from the centre of the web, a wide opaque ribbon descends in a zigzag across the radii. This is the *Epeira's* trademark, the flourish of an artist initialling her creation.

The fiery locust, who releases the spring of his long shanks at random, is particularly prone to fall into the trap. One might imagine that he would frighten the spider with his strength; the kick of his spurred levers should enable him to make a hole and get away. Yet if he does not free himself with the first effort, the locust is lost.

Turning her back on the game, the *Epeira* works all her spinnerets, pierced like the rose of a watering can, at one and the same time. The silky spray is gathered by the hind legs into a wide arc to allow the stream to spread. She flings this iridescent sheet over her prey and then, turning her prey over and over, swathes it completely.

When all movement ceases under the snowy winding sheet, the spider approaches her bound prisoner. She gnaws at the locust with her poison fangs and then withdraws, leaving the torpid victim to pine away.

Soon she comes back; she sucks the prey, drains it, repeatedly changing her point of attack. At last the clean-bled remains are flung out of the net and the spider returns to her ambush in the centre of the web.

Banded *Epeira* (*Argiope bruennichi*).

73

Other Epeirae

In my enclosure, which I have stocked carefully with the most famous breeds, I have six different species under observation, all of a useful size, all first-class spinnners. Their names are the banded *Epeira*, the silky *Epeira*, the angular *Epeira*, the pale-tinted *Epeira*, the diadem *Epeira* and the crater *Epeira*.

I am able all through the fine season to watch them at work, now this one, now that, according to the chances of the day. To describe the separate progress of the work of each of the six *Epeirae* would be uselessly repetitive: all six employ the same methods and weave similar webs, apart from certain details.

My subjects are young. The belly, the wallet containing the rope works, hardly exceeds a peppercorn in bulk. These youngsters have one precious advantage for the observer: they work by day, whereas the old ones weave only at night, at unseasonable hours.

The spinstresses of my enclosure start work in the late afternoon. It is at this time that they leave their hiding places, select their posts and begin to spin. Without any appreciable system, the spider runs about the rosemary hedge, from the tip of one branch to another, within the limits of some 45 centimetres. She comes and goes impetuously, as though at random. The result is a scanty and disordered scaffolding.

A special thread, the foundation of the real net, is then stretched across the area. It can be distinguished from the others by its isolation, its position at a distance from any twig that might interfere with its swaying length. It never fails to have, in the middle, a thick white point, formed of a little silk cushion.

The time has come to weave the hunting snare. The spider starts from the centre, which bears the white signpost, and, running along the transversal thread, reaches the circumference of the future web. Still with the same sudden movement, she rushes from the circumference to the centre. She starts again; backwards and forwards, makes for the right, the left, the top, the bottom, and always returns to the central landmark by roads that slant.

Pale-tinted *Epeira* (*Araneus pallidus*).

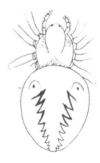

Angular *Epeira* (*Araneus angulatus*).

74

After setting a few spokes in one direction, the *Epeira* runs across to the other side to draw some in the opposite direction. These sudden changes are of course highly logical; they show us how proficient the spider is in the mechanics of rope-construction. In the end the rays are equidistant and form a beautifully regular orb. Their number is a characteristic of the different species. The angular *Epeira* places 21 in her web, the banded *Epeira* 32, the silky *Epeira* 42. These numbers are not absolutely fixed but the variation is very slight.

The laying of the radii is finished. The spider now takes her place in the centre, on the little cushion and, with an extremely thin thread, she describes from spoke to spoke a spiral line with very close coils. The thread becomes thicker as she moves outwards from the centre. Let us not be misled by the word 'spiral', which conveys the notion of a curved line. All curves are banished from the spider's work; nothing is used but the straight line and its combinations.

To this polygonal line, a work destined to disappear as the real toils are woven, I will give the name auxiliary spiral. Its object is to supply crossbars, supporting rungs.

The time has now come to work at the essential part, the snaring web for which all the rest is but a support. Clinging on the one hand to the spokes and on the other to the chords of the auxiliary spiral, the *Epeira* covers the same ground as when laying the spiral, but in the opposite direction. Ceaselessly, she turns and turns, drawing nearer to the centre and repeating the operation of fixing her thread at each spoke she crosses.

A good half hour, an hour even among the full-grown spiders, is spent on spiral circles, to the number of 50 for the web of the silky *Epeira* and 30 for those of the banded and angular *Epeirae*.

Both male and female garden spiders spin webs when they are young, but the adult male gives up catching prey in favour of courting, which can be a dangerous business as the female is much larger. Although spider silk is very elastic and very tough, the strands of the garden spider's web are no more than 0.003 mm in diameter and are regularly torn by struggles with the spider's prey or by the wind. Most garden spiders remake their webs every day.

Crater *Epeira* (*Araneus scolpetarius*).

Garden spider (*Araneus diadematus*) building web.

75

Above: the small black scorpion (*Euscorpius flavicaudis*) (2–4 cm) is common in the South of France. It lives in old buildings and prefers dampish places such as the odd corners of bathrooms and under flower pots.

Below: the larger Languedocian scorpion (*Buthus occitanus*) (6–8 cm) is found typically west of the Rhône in the departments of Gard and Hérault, although it does venture into Vaucluse where Fabre encountered it. Its sting can be dangerous!

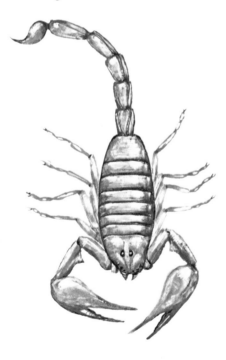

Scorpions

The common black scorpion occurs over much of southern Europe. He frequents the dark spots near our dwelling places; on rainy days in autumn, he makes his way into our houses, sometimes under our bedclothes. This weird beast causes us more fright than damage. Overrated in reputation, he is repulsive rather than dangerous. Much more to be feared and less well known is the Languedocian scorpion, which is found only around the Mediterranean region. Far from seeking our dwelling places, he keeps well out of the way. Beside the black scorpion he is a giant who, when full-grown, measures eight or nine centimetres in length. His colouring is that of pale, withered straw. It is this creature that I shall investigate more fully.

To observe their habits, I house my captives in a large glass container well provided with broken crocks to serve them as refuge. There are a couple of dozen scorpions all told. In April, a revolution takes place among my otherwise peaceful scorpions. Often under the same stone are two scorpions, the one eating the other. The devoured are always of medium size, their lighter shade and less protuberant bellies marking them as males — always males. This suggests nuptial rites tragically performed by the female after pairing.

The following spring I pay more attention to these antics. From mid-April onwards, every evening, great animation reigns within this crystal palace. The scorpions remind one of mice scuttling about. A lantern hung outside the container enables me to follow events. At times there is a brisk tumult: a confused mass of swarming legs, snapping claws, and tails curving and clashing; threatening or fondling, it is hard to say which.

I try to unravel the writhing bodies. Two scorpions, one of either sex, face each other with claws outstretched and fingers clasped. Their tails prettily curled, the couple stroll with measured steps. The male is ahead and walks backwards; the female follows obediently, clasped by her finger tips and face to face with her leader. The stroll halts occasionally, but this alters

76

nothing in the manner of the tie; it is resumed, now here, now there, from end to end of the enclosure. At times the male turns gracefully to right or left and places himself side by side with his companion. Then for a moment, with his tail laid flat, he strokes her spine. She stands motionless, impassive. For over an hour I watch, and then something happens. The male has found a shelter to his liking. He releases his companion with one hand, and, continuing to hold her with the other, he scratches out with legs and tail a shallow opening in the sand underneath one of my crocks. He enters and slowly, without violence, drags his patient partner after him. A plug of sand closes the dwelling. The couple are at home.

To disturb them now would be a blunder: I should be interfering too soon; and it does not do for me, who have turned eighty, to sit up so late. I feel my legs giving way and my eyes seem full of sand. Let us go to sleep.

All night long I dream of scorpions. They crawl under my bedclothes, they pass over my face; so many curious things do I see in my imagination. The next morning at daybreak I raise the stoneware. The female is alone.

Above: left, two scorpions fighting; right, a pair of Languedocian scorpions showing a preliminary courtship exercise known as the 'straight bend'.

Below: Languedocian scorpion killing and eating a locust.

I try again and weeks later chance favours me. A frisky, sprightly male, in his hurried rush through the crowd, suddenly finds himself face to face with a passer-by who takes his fancy. She does not say no, and things move quickly.

The foreheads touch, the claws work, the tails swing with a wide movement; they stand up vertically, hook together at the tips and softly stroke each other with a slow caress. The two animals perform the 'straight bend'. Soon, the raised bodies collapse; then with fingers clasped the couple start on their stroll. At times the male indulges in curious exercises. He draws his claws, or let us say his arms, towards himself and then stretches them straight out again, compelling the female to do likewise in an alternate fashion. The two of them form a system of joined rods, opening and closing.

After this the mechanism contracts and remains stationary. Now the foreheads touch; the mouths come together. The word 'kisses' comes to mind to express these caresses. It is not applicable, for head, face, lips, cheeks are all missing. The animal, clipped as though with pruning shears, has not even a muzzle. Where we look for a face we are confronted with a dead wall of hideous jaws, which to the scorpion doubtless represent the supreme beauty!

For a while the female lets her admirer have his way; then suddenly she brings her tail down with a bang upon the wrists of her too-ardent lover, who promptly lets go. The match is broken off for the time being. Tomorrow the sulking will be over and things will resume their course.

'In order to observe their domestic manners, I lodge my captives in a large glass volery, with big potsherds to serve as refuge.'

The Family

An anatomical monograph—the work, indeed, of a master— had told me that the Languedocian scorpion is big with young in September. Oh, how much better for me had I not consulted it! The thing happens much earlier, at least in my part of the country, the period of birth for both the Languedocian and the black scorpion being towards the end of July.

I find real eggs under the mother's belly, 30 to 40 in the case of the Languedocian scorpion. Now, how does the hatching take place? I enjoy the remarkable privilege of witnessing it. I see the mother delicately, with the point of her mandibles, seizing, lacerating, tearing off and finally swallowing the egg membrane. So the young are nicely wiped clean and free. They are white and measure but nine millimetres from tip to tail. As soon as they are free, they climb, one by one on to the mother's back, hoisting themselves along the claws which the mother scorpion keeps flat on the ground to facilitate the ascent. In order to 'cast their skin', they must remain on their mother's back for eight days without moving.

When it is done, the little ones present the normal appearance of scorpions and, having also acquired agility, they nimbly run and play around the mother. The family remains on and around the mother's back for a fortnight.

What do these young scorpions eat? The tiniest of tiny beasties—which I cannot provide. They want to go in search of it. The best thing is to say goodbye, not without a certain regret on my part. Tomorrow I will set them free on the rock-strewn slopes where the sun is hot. There they will learn the hard struggle for life better than they would with me.

'So here we have the young nicely wiped, clean and free. They are white. Their length, from the forehead to the tip of the tail, measures 9 mm. As the remnants of the egg are discarded, the young climb, first one and then the other, onto the mother's back, hoisting themselves, without excessive haste, along the claws, which the mother scorpion keeps flat on the ground, in order to facilitate the ascent.'

79

The Capricorn of the Oak

When wedge and mallet are at work preparing my provision of firewood, a favourite relaxation of mine gives me a welcome break from my daily output of prose. By my express orders, the woodman has selected the oldest and most ravaged trunks. My taste brings a smile to his lips. He wonders why I prefer wood that is worm-eaten to sound wood, which burns so much better. I have my views on the subject; and the worthy man submits to them.

Larva of capricorn beetle.

My fine oak trunks, seamed with scars, gashed with wounds, trickling with brown drops, smelling of the tanner's yard, what do your flanks contain? In the dry and hollow parts, groups of insects capable of living through the bad season of the year have taken up their winter quarters; in the live wood filled with juicy saps, the larvae of the capricorn beetle *Cerambyx miles*, chief author of the oak's undoing, have set up their home.

As insects of superior organization, these grubs are strange creatures: bits of intestines crawling about. At this time of the year, the middle of autumn, they are two different ages. The older are about a finger thick; the others hardly attain the diameter of a pencil. I find, in addition, pupae more or less fully coloured, ready to leave the trunk when the hot weather comes again. Life inside the wood therefore lasts three years. How is this long period of solitude and captivity spent? In wandering lazily through the thickness of the oak, in making roads whose rubble serves as food.

With its carpenter's gauges — strong black mandibles, short, devoid of notches, scooped into a sharp-edged spoon — the larva digs the opening of its tunnel. For the harsh work demanded of these two gauges, the larva concentrates its muscular strength in the front of its body, which swells into a pestle head. The chisels of the head are strengthened with a stout, black, horny armour that surrounds the mouth; yet, apart from its skull and its equipment of tools, the grub has a skin as fine as satin and as white as ivory.

The legs, mere rudimentary vestiges, are of no use for

80

walking. The organs of locomotion are altogether different. The first seven segments of the abdomen bristle with rough protuberances. These are the organs of locomotion, the walking surfaces. With the double support of its back and belly, with alternate puffings and shrinkings, the animal easily advances or retreats along its gallery.

Adult capricorn beetle (*Cerambyx miles*).

The cherry tree supports the small black capricorn beetle, *Cerambyx cerdo*. Beneath the tree's ragged bark, which I lift in wide strips, swarms a population of larvae all belonging to this beetle. There are big larvae and little larvae as well as accompanying nymphs. These details tell us of three years' larval life frequent in the longicorn beetles. If we hunt the thick of the trunk, it shows us not a single grub anywhere; the entire population is encamped between the bark and the wood. Here we find an inextricable maze of winding galleries crammed with packed sawdust, crossing and recrossing, shrinking into little alleys, expanding into wide spaces and cutting on the one hand into the surface layers of the sap wood and on the other into the thin sheets of the inner bark. The position speaks for itself: the larva of this beetle has tastes very different from those of its kinsman, the capricorn of the oak; for three years it gnaws the outside of the trunk, while the capricorn of the oak seeks a deeper refuge and gnaws the inside.

The dissimilarity is yet more marked in the preparation for adult life. Then the 'worm' of the cherry tree leaves the surface and penetrates into the wood to a depth of about two centimetres, leaving behind it a wide passage which is hidden on the outside by a remnant of bark. This spacious vestibule is the future insect's path of escape; the screen of bark, so easily destroyed, is the curtain that masks the exit door. In the heart of the wood, the larva scoops out the chamber destined for the nymphosis. The entrance is blocked first by a plug of fibrous sawdust, then by a chalky lid shaped like the cup of an acorn. A thick layer of fine sawdust packed into the concavity of this lid completes the barricade. Need I add that the grub lies down and goes to sleep for the nymphosis with its head against the door?

The two capricorns have the same system of closing their cells. The similarities of habit go no further. The capricorn of the oak inhabits the deep layers of the trunk; the capricorn of the cherry tree inhabits the surface. For the transformation, the first ascends from the wood to the bark, the second descends from the bark to the wood. Though the burrows have a similar form, they are produced in a reverse order. Thus we learn from the two *Cerambyx* beetles that the tool does not govern the trade.

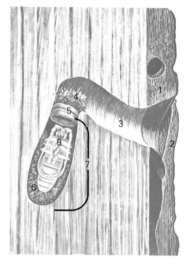

Pupation of the capricorn beetle of the cherry tree (*Cerambyx cerdo*):
1 Bark
2 The escape hatch — a loose piece of outer bark
3 Tunnel
4 Outer wood shavings
5 Chalky concave 'stopper'
6 Inner layer of fine wood shavings
7 Pupal chamber
8 Pupa

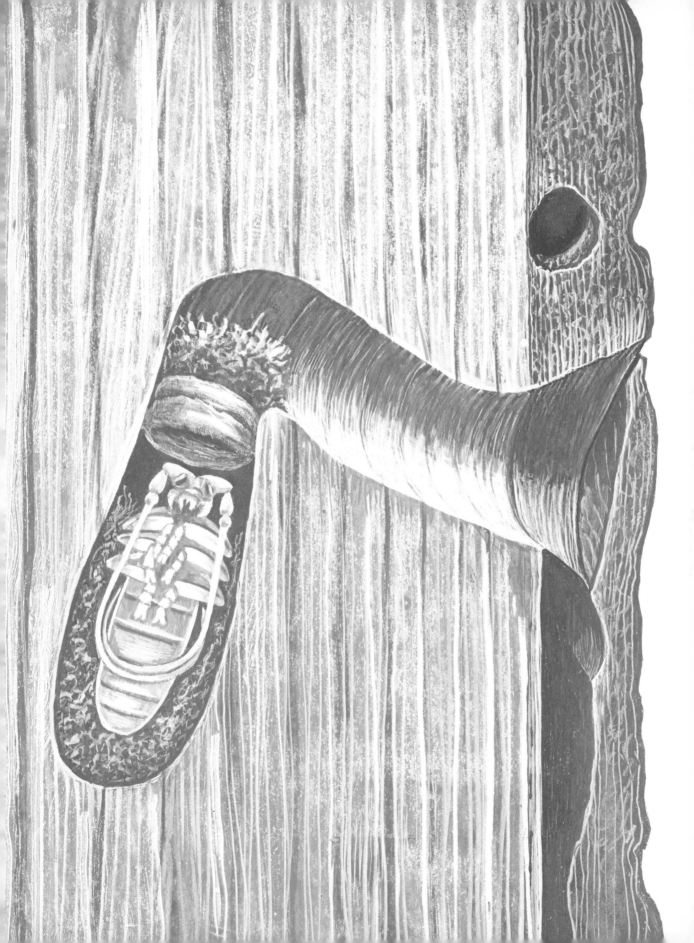

The Wood Borers

Fascinated by the habits of the two capricorn beetles, Fabre investigated the rotting trunks and bark of different trees to find out more about the behaviour of other wood-boring beetles. Old fruit trees and rotten stumps proved rich in galleries and grubs. Yet much searching revealed that the life histories of all these beetles are very similar to those of the two capricorns.

Fabre concludes his section on wood borers: 'The general law stands clearly: the wood-eating grubs of the longicorn and bupestris beetles prepare the path of deliverance for the adult, which either has to pass through a barricade of shavings deep in the wood or merely to pierce a slight thickness of bark. Thanks to the curious reversal of its usual attributes, youth is here the season of energy, of strong tools, of stubborn work; adult age is the season of leisure and idleness.'

But what of the adults that Fabre appears to discard? The attractive saperda longhorn beetles feed on the leaves of trees, especially varieties of poplar. They eat out the central part of the leaf, making holes that are characteristic for each species. Adults of other wood borers feed little and obtain moisture from tree sap or fallen fruit, a common practice among capricorn beetles. Species of Clytus can often be seen sucking the nectar from flowers. One species, the wasp beetle, Clytus arietis, common over much of Europe, visits flowers in early summer. To watch it as it crawls over a flower head, one would never think this strikingly marked beetle was the product of a wood-boring grub. Another, the musk beetle (Aromia moschata), whose larva feeds on willow, gives out a beautiful scent that bears a strong resemblance to attar of roses.

The French poet Planet, a contemporary of Fabre, eloquently describes its property:

'Il embaume, dit un naturaliste connu, tous
les alentours des saules qu'il habite,
ses émanations le trahissent fatalement
au collectionneur qui le poursuit.'

(Its scent pervades and lingers round
the clumps of willows where it lives.
By this perfume it is betrayed
to those collectors who pursue.)

1

1 *Anthexia nitidua* in cherry (*Prunus padus*).
2 *Chrysobothyrys chrysostigmata* in cherry.
3 *Arhopalus ferus* in pine (*Pinus* sp.).
4 *Bupestris octaguttata* in pine.

2

84

Fabre's account of the winding paths and burrowing grubs takes us on an almost magical journey beneath the bark and into the wood of his native trees. And one might believe from his descriptions that each type of beetle was attracted to one type of tree alone. Actually, it appears that of the beetles he describes most attack a variety of trees.

As a group, the longicorn beetles of the family Cerambycidae are serious pests of fruit trees and commercial timber. The tropics suffer most from the damage caused by the 20,000 species of the beetles so far recorded, but plantations and large orchards in Europe also experience the ravages of these beetles.

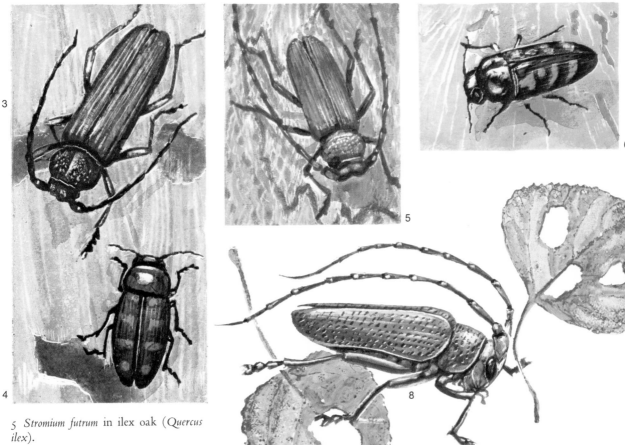

5 Stromium futrum in ilex oak (*Quercus ilex*).
6 Ptosima undeccemaculata in apricot (*Prunus armeniaca*).
7 Clytus tropicus in hawthorn (*Crategus monogyna*).
8 Saperda carcharias in black poplar (*Populus nigra*).

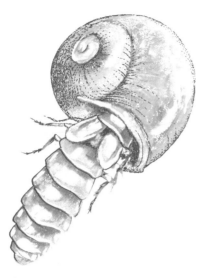

Above: the larva of the glow-worm (*Lampyris noctiluca*) is a voracious predator—feeding mainly on snails, which it reduces to a liquid form before eating.

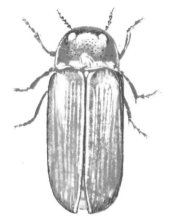

Above: the male glow-worm looks more like a typical beetle than the female and unlike her is capable of flight. Below: the underside of the male's abdomen.

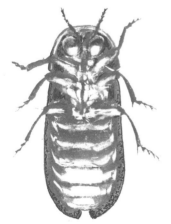

Glow-worms

This is not a worm at all, not even in general appearance. It has six short legs, which it well knows how to use; it is a gad-about, a trot-about. In the adult state, the male is correctly garbed in wing cases like the true beetle that he is. The female is an ill-favoured thing who knows nothing of the delights of flying: all her life long she retains the larval shape.

We French have the expression 'nu comme un ver' (naked as a worm) to describe the lack of any defensive covering. This creature is clothed: it wears a hardened skin; moreover, it is rather richly coloured: its body is dark brown all over, set off with pale pink on the thorax. Finally, each segment is decked at the hinder edge with two spots of a fairly bright red. A costume like this was never worn by a worm.

That master of the art of gastronomy, Brillat-Savarin, said: 'Show me what you eat and I will tell you who you are!' A similar question should be addressed to every insect whose habits we wish to study, for from the least to the greatest in the zoological progression the stomach holds sway.

Well in spite of his innocent appearance, the glow-worm is an eater of flesh, a hunter of game; and he follows his calling with rare villainy. His usual prey is a small snail hardly the size of a cherry; *Helix variabilis*, for example, which in the hot weather collects in clusters on the stiff stubble. The glow-worm briefly investigates the prey, which, according to its habit, is wholly withdrawn into the shell, except for the edge of the mantle which projects slightly. The hunter then draws his weapon, consisting of two mandibles bent back powerfully into a hook, very sharp and as thin as a hair.

It taps the snail's mantle repeatedly with its instrument. Yet it does so with such gentleness as to suggest kisses rather than bites. The first few tweaks—there are never many—are enough to impart inertia and loss of all feeling to the mollusc, which is affected by some poison from the glow-worm's tiny grooved hooks.

86

Light is produced by the oxidation of a substance called luciferin. The wingless female produces the strongest light, which is a pale greenish blue, although all stages of the life history including the eggs are bioluminescent.

Does the glow-worm divide his food piecemeal and carve it up into minute particles? I think not. He does not eat in the strict sense of the word: he drinks his fill; he feeds on the thin gruel into which he transforms his prey as does the maggot of the fly.

The lighting apparatus of the female occupies the last three segments of the abdomen. The lower surfaces of each of the first two of these segments takes the form of a wide belt; on the third segment the luminous part is much less, consisting simply of two small crescent-shaped markings which shine through to the back and are visible both above and below the animal. Belts and spots emit a glorious white light, delicately tinged with blue.

The two belts, an attribute exclusive to the marriageable female, form the part richest in light. These beacons are evidently nuptial signals and to aid their effect the females indulge in violent exercises; twist the tips of their abdomens, turn them to one side, turn them to the other, jerk them in every direction. In this way the searchlight cannot fail to gleam before the eyes of every male who goes a-wooing in the neighbourhood.

The male, on his side, is provided with an optical apparatus able to catch the least reflection. His corselet expands into a shield and overlaps his head considerably in the form of a peaked cap or eye-shade. Beneath this arch are two eyes, which are relatively enormous.

From start to finish the glow-worm's life is one great orgy of light. The eggs are luminous; the grubs likewise. We can understand the object of the female's beacon; but of what use is all the rest of this pyrotechnic display? To my great regret, I cannot tell. It is, and will remain for many a day to come, perhaps for all time, a secret of animal physics.

1 Scarab beetles (*Scarabeus sacer*) arrive at a fresh pile of dung.
2 The edge of the broad flat head is used to dig and cut up the dung. The front legs armed with five strong teeth serve as rakes.
3 The selected dung is pushed under the beetle's belly between the hind legs which are long and used to fashion the dung into a ball. The ball is eventually pushed away, the hind-legs acting as pivots.

The Sacred Scarab — Dung Roller

We started off one morning down a path fringed with dwarf elder and hawthorn, whose clustering blossoms were already a paradise for the rose-chafer. We were going to see whether the sacred beetle had yet made his appearance on the sandy plateau of Les Angles.

The time was right, the scavengers were at work — those beetles whose mission it is to purge the soil of its filth. One can never weary of admiring the variety of tools with which they are supplied for sifting, cutting up and shaping the dung, or for excavating the deep burrows in which they seclude themselves with their booty.

What excitement over a single patch of dung! Before the sun becomes too hot, the creatures are there in their hundreds, large and small of every sort, shape and size, hastening to carve up the common cake.

Who is this arriving at the heap? His long legs move with awkward jerks, his little red antennae unfurl their fan. He is coming, he has come, not without sending a few banqueters sprawling. It is the sacred beetle, clad all in black, the biggest and most famous of our dung beetles.

Let us watch the construction of the famous ball. The clypeus, the edge of the broad flat head, is notched with six angular teeth arranged in a semicircle. This is the digging and cutting tool, and the rake that lifts and casts aside the unnutritious vegetable fibres. The fore-legs play a large part in the work. They are flat bow-shaped, supplied on the outside with five strong teeth. If an obstacle has to be removed from the thickest part of the heap, the scarab uses his elbows; that is to say he flings his toothed legs to the right and left with an energetic sweep, and clears a semicircular space. Once room is made, these fore-limbs collect armfuls of the stuff raked together by the head, and push it under the insect's belly, between the four hind-legs. These are long and slender, especially the last pair, slightly bowed and finished with a very sharp claw. They act like compasses, capable of embracing a globular body in their curved branches — their function is to fashion the ball.

Armful by armful, the material is heaped together under the belly, between the four legs, which by slight pressure give it a

4 The scarab does not always push his ball alone; sometimes he takes a partner or to be more accurate the partner takes him. The newcomer goes through the motions of helping, usually stationed at the front.

89

curve and a preliminary outline. Every now and then the rough-hewn pill is sent spinning between the back legs and in this way is turned under the dung beetle's belly until it is rolled into a perfect ball.

Under an oppressively hot sun, one stands amazed by the beetle's feverish activity. The work proceeds: that which a moment ago was a tiny pellet is now a ball the size of a walnut; soon it will be the size of an apple. I have seen some gluttons manufacture a ball the size of a man's fist.

The beetle has his provisions. Now he must withdraw from the crowd and transport his food ball to a quieter spot. Here the scarab's true character shows itself. Clasping his sphere with his two hind-legs, whose terminal claws serve as pivots, he grips the mass with the middle pair of legs; and with his toothed fore-arms pressing in turn upon the ground, he proceeds with his load. The rear legs are in continual movement backwards and forwards, the claws shifting to change the axis of rotation. In this way the ball touches the ground by turns at every point of its surface, a process which perfects its shape and gives consistency to the outer layers.

The burrow itself is a shallow cavity, roughly the size of a man's fist; dug in the soft earth, it communicates with the outside by a short passage just wide enough to admit the ball. As soon as the provisions are safely stored away, the scarab shuts himself in by stopping up the entrance with rubbish kept in the corner for the purpose. Once the door is closed, nothing outside

5 The original dung beetle finds a place to dig a burrow. The work proceeds quickly and every time it resurfaces it checks to see that its precious ball of dung is still there. Sometimes it will move it nearer the burrow entrance.

6 As the burrow gets deeper the excavator appears less at the surface—now is the time for the thief to act. He grips the ball and speedily makes off with it.

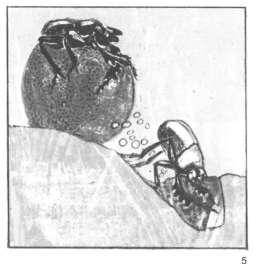

5 6

betrays the existence of the banqueting chamber. The table is sumptuously spread; the ceiling tempers the heat of the sun and allows only a moist and gentle warmth to penetrate. So complete is the illusion, I think of the saying:

Ah! qu'il est doux de ne rien faire
Quand tout s'agite autour de nous?

(Ah, how sweet to have nothing to do
When all around us throbs the busy world.)

The Sacred Scarab — its Feeding

The ball fills almost the whole room; the rich feast rises from floor to ceiling. Here sits the banqueter, belly to the table, back to the wall. The great ball of dung passes mouthful by mouthful through the beetle's digestive canals, yielding up its nutritive essences, and reappears at the opposite end spun into an unbroken cord — ample proof that digestion goes on uninterrupted. When the whole ball has passed through the machine, the hermit returns to the daylight to try his luck afresh, finds another patch of dung, fashions a new ball and starts eating again. This life of pleasure last for a month or two from May to June. Then, with the coming of the fierce heat, the sacred beetles take up their summer quarters and bury themselves in the cool earth. They reappear in the autumn, less numerous, less active but now seemingly absorbed in the most important work of all — the perpetuation of the species.

7 The owner usually knows what happened and pursues the thief who more often than not will act 'dumb' and pretend he was just retrieving the ball that happened to roll away. Sometimes there is a fight.

8 Often the whole story ends happily enough, the ball of dung is rolled into the newly made burrow and both beetles spend many days digesting the dung, side by side in the security of their underground retreat.

7   8

The Spanish Copris—Dung Hoarder

The Spanish copris (*Copris hispanus*) is round and squat with a ponderous gait, and it knows nothing of the gymnastics of the scarabs. Its legs, which are of insignificant length, bear no comparison with the stilts of the pill-rollers.

The copris is indeed of sedentary habit. Once it has found its provisions at night or in the evening twilight, it digs a burrow under the heap, a rough cavern. Here it introduces, bit by bit, the stuff that is just overhead. As long as the hoard lasts, the copris, engrossed in the pleasure of the table, does not return to the surface. The home is not abandoned until the larder is emptied; then the insect recommences its nocturnal quest, finds new treasure and scoops out another temporary dwelling.

Female copris (*Copris hispanus*) in the underground chamber with her loaf of dung. Top left: the male retiring.

92

But in May, or June at latest, comes egg-laying time. The insect, so ready to fill its own belly with the most sordid materials, becomes particular when the family is concerned. Like the sacred beetle, it now wants the soft produce of sheep, deposited in a single slab. However abundant it may be, the cake is buried on the sport in its entirety. Not a trace of it remains outside.

The work continues for the best part of a night. During the days that follow nothing happens; the copris goes out no more. Before the week is out, I dig up the soil. The burrow is a spacious hall with an irregular elliptical roof and an almost level floor. In a corner is a small round hole; this is the goods entrance, opening on a slanting gallery that runs up to the surface of the soil. The ordinary eating chamber may be a hole hurriedly scooped out, but this dwelling is of larger dimensions and much more carefully built.

I suspect that both sexes have a share in this architectural masterpiece; at least, I often come upon the pair in the burrows destined for the laying of the eggs. But once the home is stocked, the male retires discreetly to the outside world.

Inside the burrow, I always find a single lump of dung, a huge loaf which fills the dwelling except for a narrow passage all around just wide enough to give the mother room to move.

This sumptuous portion, a regular Twelfth-night cake, has no fixed shape. I come across some that are ovoid like a turkey's egg, some like a common onion and some that remind me of a Dutch cheese. In all cases the surface is smooth and nicely curved.

The mother has collected and kneaded into one lump the numerous fragments brought down one after the other; out of all these particles she has made a homogeneous thing by mashing them, working them together and treading on them.

When the baker has kneaded his dough to the right extent, he collects it into a single lump in a corner of his baker's trough. The copris knows this bakehouse secret. She heaps together all that she has collected, carefully kneads it into a provisional loaf and allows it time — a week at least — to improve, to give flavour to the paste and make it of the right consistency for subsequent manipulations.

At last it is ready. The baker's man divides his lump into smaller lumps, each of which will become a loaf. The copris does the same thing. By means of a circular cut made by the sharp edge of her forehead and the saw of her fore-legs, she detaches from the mass a piece of the prescribed size. With one sharp, decisive cut, she obtains the proper-sized lump.

There now remains the job of shaping it. Clasping it in her short arms, so little adapted, one would think, to work of this kind, the copris rounds her lump of dough by means of pressure, and of pressure alone. After twenty-four hours of this, the piece that was all corners has become a perfect sphere the size of a plum.

For a long time the insect continues to touch up its globe, polishing it affectionately. These meticulous finishing touches seem endless. Towards the end of the second day all is finished. The mother climbs to the dome of her edifice and there, still by simple pressure, hollows out a shallow crater. In this basin the egg is laid.

Then, with extreme caution, the lips of the crater are brought together to form a vaulted roof over the egg. All's well, it seems; and once again she resumes her patient toil: the careful, delicate scraping of the sides towards the summit. Twenty-four hours more are spent in this minute work. Total: four times round the clock and sometimes longer to construct the sphere, scoop out the basin, lay the egg and shut it in.

The insect goes back to cut the loaf, help itself to a second slice, which by the same manipulations as before becomes an ovoid tenanted by an egg. The surplus suffices for a third ovoid, sometimes a fourth.

The laying is over. Here is the mother in her retreat, which is almost filled by the cradles standing one against the other, pointed end upwards. What will she do now? Go away to recruit her strength? Not a bit of it; she stays. The copris is a devoted mother, who braves hunger rather than let her offspring suffer privation.

Assiduously she goes from one to the other; feels them, listens to them, touches them, remedying any accident no matter how trifling. Under her never-failing vigilance the pill does not crack, for any crevice is stopped up as soon as it appears; nor does it become covered with parasitic vegetation, for nothing can grow on a soil that is constantly raked.

Three months later, in September, mother and progeny emerge at the time of the first autumnal rains. Together they arrive at the autumnal banquets, when the sun is mild and the ovine manna abounds along the paths.

From a great loaf of dung she buries underground, the female Spanish copris manufactures three or four spherical plum-sized cakes. In each of these she lays a single egg. She is noted for her extreme maternal care — the cakes are constantly scraped and cleaned to prevent fungus and mildew growing on them. She stays underground from June when the eggs are laid until well into October when the young hatch out.

Insect Colouring

Phanaeus splendidulus, the glittering, the resplendent — this is the name selected by the scientists to describe the handsomest dung-beetle of the pampas. The name is not at all exaggerated. Combining the fire of gems with metallic lustre, the insect emits the green reflections of the emerald or the gleam of ruddy copper, according to the incidence of light. This muck-raker would do honour to the jeweller's show-cases.

Our own dung-beetles, though usually modest in their attire, also have a leaning toward luxurious ornament. One, *Onthophagus*, decorates his corselet with Florentine bronze; another wears garnets on his wing-cases. Black in all parts exposed to the light of day, the *Sterocarius geotropes* displays a lower surface of a glorious amethyst violet.

Many other beetles of greatly varied habits even surpass the dung-beetles in the matter of jewellery. For example the azure beetle *Hoplia caerula*, an inmate of the osier-beds and elders by mountain streams, is a wonderful blue, tenderer and softer to the eye than the azure of the heavens. You could not find an ornament to match it save on the throats of certain humming-birds and the wings of a few butterflies of the tropics.

To adorn itself like this, by what bonanza does the insect gather its gems? What a pretty problem is that of a *Bupestris* beetle's wing case! Here the chemistry of colours ought to reap a delightful harvest. But the difficulties are great, so great, it seems, that science cannot yet tell us the why and wherefore of the humblest costume.

Phanaeus splendidulus

Sterocarius geotropes

Hoplia caerula

96

Fabre's observations on the colouring of insects dates back to his first studies of the hunting wasps, when he noticed that their skin changed from transparent to a white colour. He dissected the larvae and found the skin to be made up of two kinds of vesicles: the one, yellow and transparent, containing the normal fatty tissues; the other, fine white particles that imparted to the larva its overall white colour. He dissolved the particles in nitric acid, heated the solution and to the residue added ammonia, causing a 'glorious crimson colour'. From this chemical reaction he deduced that the particles consist of uric acid or, more precisely, ammonium urate — a common excretory product of insects. He performed experiments on a number of subjects including the decticus locust and the beautifully marked spurge moth caterpillar; in all cases he found that the coloured parts readily dissolved in acid and upon the addition of ammonia produced the typical red murexide reaction.

This still left the problem of his glorious beetles. He knew that iridescence must be caused in some way by a physical property of the insect's hardened skin, the cuticle, but with his crude techniques he was not able to investigate further. He could only deduce that beetles and other insects absorbed their own waste products to produce their colourful coats. On the brilliant colours of insects and the higher animals, he concluded, 'What are they in reality? Answer: drops of urine.'

Whilst correct in his belief that uric acid plays a role in colouring, Fabre held an over-simplified view of the intricacies involved. Insect coloration can be divided into two main types: colours that are formed by the structural arrangement of the insect's cuticle, and pigments of various chemical compositions.

The iridescence of beetles is due to the complicated arrangement of plates or lamellae in the cuticle, which causes the optical interference of certain wavelengths of incident light. The spacing between these plates and the presence or absence of a black backing pigment in the inner layer of the cuticle produce the typical metallic sheens of dung and predatory beetles.

Butterflies, bugs and certain other beetles such as ladybirds owe their colour to pigments. These are coloured because they absorb light of certain wavelengths and reflect others. Our eyes perceive the colour complementary to that which has been absorbed. Insects may use structural coloration combined with a mixture of different pigments.

Minotaurus Typhoeus

Below: vertical section of *Minotaurus typhoeus* burrow.
Right: vertical section enlarged to show bottom of burrow. The male grinds a sheep dropping up by spearing it with his three prongs; holding it fast, he saws off pieces with his fore-arms. The bits fall down the burrow to the female who grinds them up into finer pieces, pats them down, then stamps on them. The egg is laid at the very bottom of the burrow underneath a conical cake of pulverized dung.

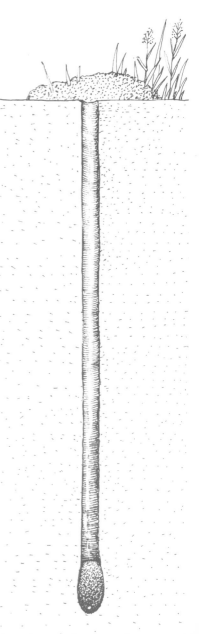

To describe this insect, scientific nomenclature joins two formidable names: that of the Minotaur, Minos' bull fed on human flesh in the crypts of the Cretan labyrinth; and that of Typhoeus, one of the giants who tried to scale the heavens.

The Typhoeus of Greek legend had the ambition to sack the home of the gods by stacking one upon the other a pile of mountains; the Typhoeus of the naturalists does not climb, he descends to enormous depths.

This is a creature of open sandy places where, on their way to the grazing ground, the flocks of sheep scatter their trails of pellets which constitute its regulation fare.

To discover its world is no easy undertaking for which the point of a knife will suffice. We have to dig out a pit, the bottom of which can only be reached with a spade, sturdily wielded for hours at a stretch. And, if the sun is at all hot, one returns from the drudgery utterly exhausted. Oh, my poor joints, grown rusty with age! Luckily I have an assistant, in the shape of my son Paul, who lends me the vigour of his wrists and suppleness of his loins. The rest of the family comes too, including my wife—and she is not the least eager. One cannot have too many eyes when the pit becomes deep and one has to observe at a distance the minute documents exhumed by the spade.

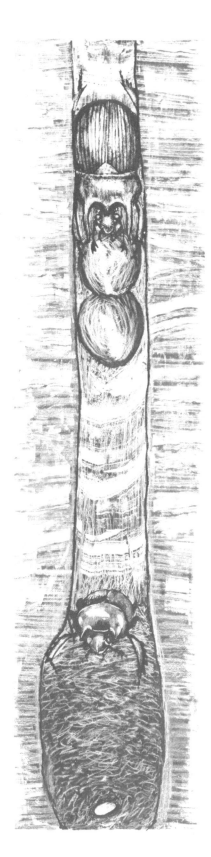

Once the pocket trowel has prudently laid things bare, we see the occupants appear: the male first and, a little lower, the female. When the couple are removed a dark circular patch appears; this is the end of the food column. Careful now and let us dig gently! What we have to do is to cut away the central clod and extract the block all in a lump. There! That's done it. We possess the couple and their nest. A morning of arduous digging has procured us these treasures.

After viewing the Minotaurus in the field and in artificial burrows at home, Fabre gives us a resumé of the life of this extraordinary dung beetle.

When the great colds are over, he sets out in quest of a mate, buries himself and her and thenceforth remains faithful to her, despite his frequent excursions out of doors. With indefatigable zeal, he assists the burrower, who is herself destined never to leave her home until the emancipation of the family. For a month or more, he loads the rubbish of excavation on his forked hod leaving the relatively easy work of the excavating rake to the mother. Once the burrow is dug the navvy turns himself into dung collector. Some distance from the bottom of the shaft, he crumbles the dung hardened by the sun, he makes it into a semolina and flour that rains down onto the maternal bakery. At last, worn out by his efforts, he leaves the house and goes out to die outside. The mother, for her part allows nothing to divert her from her housekeeping. Throughout her working life, never leaving her home, she kneads her cylindrical loaves, fills them with an egg, watches over them until the moment for the exodus arrives. Then in the autumn she accompanies her young who disperse at once to feast in the regions frequented by the sheep. Thereupon, having nothing left to do, the devoted mother perishes.

The turning point in the study of dung beetles came with the publication of Volumes 5 and 6 of Fabre's Souvenir entomologiques. *Until then descriptions had been of historical interest only. Fabre spent over forty years studying these beetles and his writings reflect his intense passion for these buriers of dung. In these studies he attains a new maturity of style in which a richness of language and freedom of expression is combined with extremely accurate observation.*

Glossary

Arachnid Class of invertebrate animals with four pairs of jointed walking legs—including spiders and scorpions. Unlike the insects, the front part of the body is never divided into head and thorax, but like the insects it bears prehensile and sensory appendages.

Aveyron Department in south-eastern France.

Brillat-Savarin, Anthelme (1755–1826), French gastronomist and writer, author of *Physiologie du goût (The Physiology of Taste)*.

Caterpillar The larval stage of a moth or butterfly, wormlike in form, and with a body consisting of 13 segments plus the head. There are 6 pairs of true legs on the thoracic segments, and also a varying number of pro-legs on the abdominal segments.

Cerceris Genus of hunting wasps, distinguished from all others by the deep constrictions between the abdominal segments.

Cocoon A case made partly or completely from silk, designed to protect the pupa. It is constructed by the larval stage of the insect.

Coleoptera A very large order of insects (with more than 250,000 species). The front wings are hard or leathery, and meet in mid-line. These wing cases or 'elytra' protect the membranous hind-wings and the vulnerable abdomen. They have biting mouthparts. To this order belong the beetles and weevils.

Dictyoptera Insect order which includes cockroaches and mantises. They have biting and chewing mouthparts, and their bodies often exhibit a flattened appearance. Long spiky legs, long antennae and well-developed anal cerci (paired appendages at the end of the abdomen).

Dufour Léon (1780–1865), an army surgeon who later practised as a doctor in the Landes region of France. Amongst several natural history studies, he wrote a monograph on *Cerceris*, the hunting wasp.

Ganglion A mass of nerve cells and nerve fibres.

100

Harmas Provençal word for a piece of stony waste ground once cultivated.

Hemiptera An order of insects, including 'bugs' and cicadas, of various shapes and sizes, but all with mouthparts adapted for sucking the juices of plants and animals. Front wings overlap each other, distinguishing them from beetles (Coleoptera).

Hunting wasps Popular name for the many species of solitary (non-social) wasps.

Hymenoptera Large order of insects (over 100,000 species), which include wasps, bees and ants. Most have a definite 'waist' between the first and second segments of the abdomen. Usually two pairs of membranous wings.

Larva The stage in the life cycle of insects between the egg and the pupa.

Lepidoptera Order of insects (moths and butterflies) whose members have a pair of membranous wings covered with small scales. Adults feed by sucking liquids, such as nectar, through a proboscis. The larvae have biting mouthparts and feed on plants. Though there is no clear-cut distinction, moths are most commonly distinguished from butterflies by their unclubbed antennae. Butterflies are all day fliers; moths may fly by day or night.

Mandible The first pair of insect mouthparts — the jaws. Often sharply toothed and used for biting in the grasshoppers and wasps.

Mantle The soft part of a mollusc (e.g., snail) which covers the main body organs.

Meloidae Family of beetles, called oil beetles. All have special chemical components in their blood, which can have blistering effects or produce foul smells. Their life history is peculiarly complicated, as Fabre discovered.

Nymph Refers to the immature form of an insect when it bears a resemblance to the adult (as in dragonflies).

Ootheca Packet of eggs bound together by secretions produced by glands in the female insect's genital ducts.

Orthoptera Order of medium to large insects, including grasshoppers, crickets and bush crickets. They have stout bodies, blunt heads and legs which are often modified for jumping. The first segment of the thorax is large and saddle-shaped.

Ovipositor The egg-laying apparatus of an insect. In some groups, such as grasshoppers and crickets, it can be extremely long.

Provençal The language of Provence, also known as *langue d'oc*. In the sixteenth century it largely died out as the common language for southern France, and it was not until the nineteenth century, under the influence of Mistral (q.v.) and others that it was revived as a literary medium.

Pupa The third stage in the life history of insects such as beetles and butterflies. It is a non-feeding, immobile stage. Within the casing of the pupa the insect's body undergoes the complete transformation which turns it into an adult.

Réaumur French scientist (1683–1757), noted for his studies of insects and for his invention of the Réaumur thermometer scale, in which freezing point is 0° and boiling point 80°.

Rodez The capital town of the department of Aveyron.

Rostrum A beak or snout; used to describe the piercing mouthparts of bugs.

Spinnerets Silk-producing glands of spiders. Up to three pairs are situated in the posterior part of the abdomen.

Stemma (plural *stemmata*) A simple eye.

Vaucluse Mountain range in south-east France; also the name of the department whose capital is Avignon.

Ventoux The highest mountain in Vaucluse (1,912 metres). The present forest dates from 1860, the mountain having been denuded in the sixteenth century. Forms a climatic barrier and sports many rare, endemic plant and insect species.

Virgil Roman poet (70–19 BC), whose writing on the countryside is mostly contained in the *Eclogues* (10 pastoral poems) and the *Georgics* ('Art of Husbandry').

Index

Roman numbers refer to the text, italic numbers refer to the illustrations and captions. Latin names used by Fabre are set in brackets wherever they differ from the modern scientific binomial. Insect measurements given in brackets after page numbers are reckoned from head to thorax, with the exception of butterflies and moths, which are measured by wing span. These sizes are average in Provence but they may differ in other regions.

Fabre observing insects. *(Collection Kurt Guggenheim)*

Bibliography

All the books by J. H. Fabre with the exception of Social Life in
the Insect World *have been translated by Alexander Teixeira De
Mattos F.Z.S. and Bernard Miall.*

Brangham, A. N., *The Naturalist's Riviera*, Phoenix House
Ltd, London, 1962

Doorly, Eleanor, *The Insect Man*, William Heinemann,
London, 1959

Fabre, J. H., *The Life and Love of the Insect*, Adam and
Charles Black, London, 1911

Fabre, J. H., *The Life of the Spider*, Hodder and Stoughton, London, 1912

Fabre, J. H., *Bramble Bees and Others*, Hodder and Stoughton, London, 1915

Fabre, J. H., *The Hunting Wasps*, Hodder and Stoughton, London, 1916

Fabre, J. H., *The Life of the Caterpillar*, Hodder and Stoughton, London, 1916

Fabre, J. H., *The Life of the Grasshopper*, Hodder and Stoughton, 1917

Fabre, J. H., *The Glow Worm and Other Beetles*, Hodder and Stoughton, London, 1919

Fabre, J. H., *The Mason Wasp*, Hodder and Stoughton, London, 1919

Fabre, J. H., *The Sacred Beetle and Others*, Hodder and Stoughton, London, 1919

Fabre, J. H., *The Story Book of Science*, Hodder and Stoughton, London, 1919

Fabre, J. H., *Social Life in the Insect World*, translated by Bernard Miall, Penguin Books, London, 1937

Guggenheim, Kurt, *Sandkorn für Sandkorn, Die Begegnung mit J. H. Fabre*, Huber Verlag, Frauenfeld, 1979

Guggenheim, Kurt and Portman, Adolf, *Jean Henri Fabre Das offenbare Geheimnis*, Insel Taschenbuch, 1977

Harant, H., and Jarry, D., *Guide du Naturaliste dans le Midi de la France*, Vol. 2, Delachaux et Niestlé, Neuchâtel, Switzerland, 1973

Legros, Dr C. V., *Fabre Poet of Science*, Fisher and Unwin, London, 1921

Les publications de l'entomologiste J. H. Fabre (1823–1915), Mason & Cie, 120 boulevard St Germain, Paris 6e, 1974

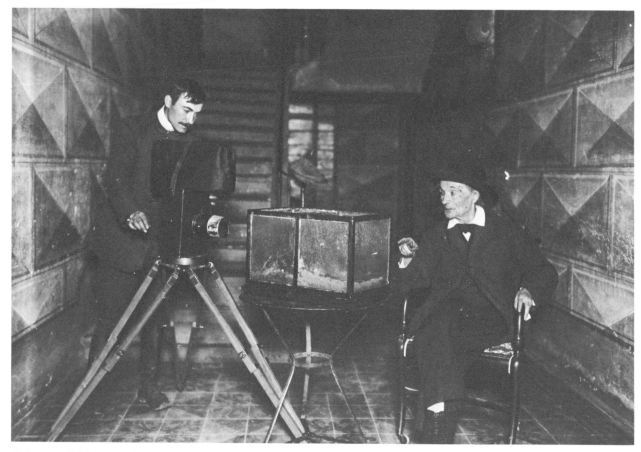

Fabre and his son photographing insects. *(Collection Kurt Guggenheim)*

les variations autour du motif de laquelle la spirale [illegible] [illegible], les [illegible]
de l'époque récente (~~actuelle~~) y sont-elles arrivées par les perfectionnements graduels d'un passé
insuffisamment?
antique? Non; dès les premiers âges du monde, la spirale logarithmique présida à
l'architecture des coquilles, les Nummulites, les Ammonites et autres races antérieures à
l'apparition de nos continents, avec journellement de la même [illegible] que le font les planètes
de nos jours. La spirale logarithmique est plus vieille que ~~aucune~~ [illegible]
vient de cette géométrie souveraine qui régente le monde, attentive à la cellule de
la guêpe comme à la rampe de l'escargot. Dans ses ouvrages le Potier céleste
toujours géométrise. Ἀεὶ ὁ θεὸς γεωμετρεῖ disait Platon. Là vraiment est la
solution du problème de la Guêpe.

Lei Lani

La Guêpe commune.

Avec des recherches assidues, la découverte de quelques nids de la Guêpe commune
(Vespa vulgaris) est assurée dans mon voisinage. En septembre je me mets à
l'aventure, inspectant du regard le bord des sentiers. Mon fils Paul m'accompagne
et me prête ses bons yeux de l'enfance, son attention que ne troublent pas encore les
tracas de la vie. ~~[illegible]~~ [illegible] vingt pas de distance, il voit de suite s'élever de
terre et disparaître des points indécis, comme si d'un petit cratère en éruption partaient
[illegible] [illegible] des projectiles. En même temps il en arrive d'autres qui s'abattent
au point même d'où les premiers sont partis. « Un nid! fait Paul, tout ému et un
[illegible] [illegible]; un nid! bien sûr. — On s'approche discrètement comme si [illegible] l'attention
[illegible] est au effet un guêpier et l'entrée du vestibule, ouverture [illegible]
ou l'épaisseur du terrain. C'est [illegible] la pomme, [illegible] [illegible] les [illegible]. Le [illegible]
où pourraient s'engager la [illegible] [illegible] [illegible] [illegible] [illegible] s'attaquer à ces
[illegible] [illegible] de trop près. Sans autres informations, qui pourraient avoir des
[illegible] [illegible] de loin. Nous reviendrons à la nuit close,
[illegible] [illegible] toute la population sera rentrée des champs.

La conquête d'un nid de Guêpes serait opération de quelque gravité si l'on n'y mettait
beaucoup de prudence. Un quart de litre d'essence de pétrole, un bout de roseau d'une
[illegible] [illegible] de longueur, un petit tampon de terre argileuse pétrie à l'avance, tels
sont les moyens d'attaque jugés les plus simples et les meilleurs après [illegible]
[illegible] ~~[illegible]~~ l'essai. La méthode par l'asphyxie au moyen du soufre;
[illegible] [illegible] [illegible] à moins ~~[illegible]~~ d'expédients coûteux [illegible]
[illegible] [illegible] [illegible]. Le bon Réaumur, quand il voulait mettre un [illegible]
[illegible] [illegible] [illegible] le [illegible], avait mis [illegible] [illegible] [illegible] qui [illegible]
un guêpier [illegible] [illegible] sans ~~[illegible]~~ [illegible] [illegible], [illegible] [illegible]
par une [illegible] d'argent [illegible]. Moi, qui devrais payer directement de ma
[illegible] la [illegible] du savant, [illegible] [illegible] [illegible] [illegible], [illegible] [illegible]
l'y [illegible] à [illegible] [illegible] avant d'[illegible] [illegible]. Au préalable, [illegible]
les habitants par asphyxie. Le moyen est brutal, mais il me donne pleine
sécurité. — D'ailleurs je n'ai pas à [illegible] ce que le [illegible] a [illegible] [illegible];
mes ambitions se bornent à quelques points de détail qu'il [illegible] [illegible] [illegible]
sur un petit nombre de [illegible]. Ces [illegible], je les [illegible] si je [illegible]
la dose de liquide asphyxiant. Je donne la préférence à l'essence de pétrole à
cause de son bas prix. Il s'agit de l'introduire sans perte dans la cavité
[illegible] [illegible] [illegible]. Un vestibule d'un empan [illegible] [illegible]